Rectifying Climate Injustice

This book provides an account of how rectificatory justice for climate change loss and damage can be realized by bridging the worlds of political philosophy, climate science and climate policy together. The book focuses on three fundamental questions: what kinds of climate impacts should count as loss and damage, how climate science can help us identify them and who should bear the burdens of providing reparations for loss and damage.

Laura García-Portela argues that loss and damage occur after people's capabilities have fallen below a threshold of sufficiency due to the negative impacts of climate change, thereby infringing people's human rights. She argues for a historical responsibility principle for reparations for loss and damage (the Polluter Pays Principle, PPP) grounded in her Continuity Account. According to this account, responsibility for reparations is based on the duty to refrain from emissions-generating activities that would infringe people's human rights. A new duty to provide reparations arises when human rights are infringed by climate change-inducing activities. Importantly, she examines how the latest developments in attribution science can help in developing a rectificatory account for loss and damage, an approach that has not been considered in depth by climate justice scholars so far.

Striving to improve the reader's understanding of loss and damage as outlined by The United Nations Framework Convention on Climate Change, this book will be of great interest to students and scholars of climate justice, environmental justice, and environmental ethics.

Laura García-Portela is an assistant professor of philosophy at Erasmus University Rotterdam. Before that, she held postdoctoral positions at the Karlsruhe Institute of Technology (KIT, Germany) and the University of Fribourg (Switzerland). She has also held visiting, research, and teaching positions at University of Valencia (Spain), Keele University (UK), and University of Washington (US). She graduated in summer 2021 from the Department of Philosophy and the Interdisciplinary Doctoral Program in Climate Change, University of Graz. Her dissertation was awarded the Luis Díez del Corral Prize from the Center of Political and Constitutional Studies in Spain (research centered attached to the Spanish Ministry of Presidency) and the Roland Atefie Preis from the Austrian Academy of Science. Her work lies at the intersection between political philosophy, philosophy of climate science, and philosophy of climate law and is published in numerous international journals.

Routledge Studies in Environmental Justice

This series is theoretically and geographically broad in scope, seeking to explore the emerging debates, controversies and practical solutions within Environmental Justice, from around the globe. It offers cutting-edge perspectives at both a local and global scale, engaging with topics such as climate justice, water governance, air pollution, waste management, environmental crime, and the various intersections of the field with related disciplines.

The *Routledge Studies in Environmental Justice* series welcomes submissions that combine strong academic theory with practical applications, and as such is relevant to a global readership of students, researchers, policy-makers, practitioners and activists.

Intergenerational Challenges and Climate Justice
Setting the Scope of Our Obligations
Livia Ester Luzzatto

Solar Technology and Global Environmental Justice
The Vision and the Reality
Andreas Roos

Environmental Justice in Early Victorian Literature
Adrian Tait

Guerrilla Ecologies
Green Capital, Nature, and the Politics of Catastrophe
John Maerhofer

Rectifying Climate Injustice
Reparations for Loss and Damage
Laura García-Portela

For more information about this series, please visit: www.routledge.com/Routledge-Studies-in-Environmental-Justice/book-series/EJS

Rectifying Climate Injustice

Reparations for Loss and Damage

Laura García-Portela

Routledge
Taylor & Francis Group
LONDON AND NEW YORK

earthscan
from Routledge

First published 2025
by Routledge
4 Park Square, Milton Park, Abingdon, Oxon OX14 4RN

and by Routledge
605 Third Avenue, New York, NY 10158

Routledge is an imprint of the Taylor & Francis Group, an informa business

British Library Cataloguing-in-Publication Data
A catalogue record for this book is available from the British Library

ISBN: 978-1-032-50834-4 (hbk)
ISBN: 978-1-032-50835-1 (pbk)
ISBN: 978-1-003-39988-9 (ebk)

DOI: 10.4324/9781003399889

Typeset in Times New Roman
by Apex CoVantage, LLC

The electronic version of this book was funded to publish Open Access through Taylor & Francis' Pledge to Open, a collaborative funding open access books initiative. The full list of pledging institutions can be found on the Taylor & Francis Pledge to Open webpage.

To my parents,
for their support in my endeavor
to become a philosopher

Contents

Acknowledgments *x*
List of Abbreviations *xii*

1 Introduction 1
1.1 An international mechanism for loss and damage 3
1.1.1 A brief history of loss and damage 3
1.1.2 Contributing to an international mechanism for loss and damage 6
1.1.3 The contested role of compensation in climate negotiations 6
1.2 A terminological note 7
1.3 Philosophical methodology 10
1.3.1 General methodology: reflective equilibrium 10
1.3.2 Climate justice specific methodology 11
1.4 Assumptions and limitations 14
1.4.1 The scope of justice: recipients and duty-bearers of climate justice 14
1.4.2 Sufficientarianism: background theory of distributive justice 19
1.5 Summary of chapters 21

2 A minimal capabilities-based approach 28
2.1 Life disruptions as harm and the minimal understanding of loss and damage 29
2.2 A minimal capabilities-based account of loss and damage 31
2.3 Answering some challenges to an ex post categorization of L&D 34
2.4 Conceptual clarifications and types of reparation for loss and damage 37

2.4.1 The notions of 'loss' and 'damage' in loss and damage 37
2.4.2 Reparations for economic damage, non-economic losses, and non-economic damage 38
2.5 Conclusion 44

3 In search for a justified rectificatory justice principle 49
3.1 Two objections against the polluter pays principle 50
3.2 The beneficiary pays principle and some intuitive reactions to the objections 51
3.3 The causation objection and the beneficiary pays principle 53
3.4 The excusable ignorance objection and fairness considerations 55
3.4.1 A fresh look at the excusable ignorance objection 55
3.4.2 Fairness considerations, the beneficiary pays principle, and replies to some objections 58
3.5 Conclusion 60

4 Reasons awaiting satisfaction 66
4.1 The continuity thesis and the continuity account 67
4.2 The continuity account and some objections against the polluter pays principle 71
4.3 Alternative accounts 73
4.3.1 Strict liability 73
4.3.2 Counterfactual liability 75
4.3.3 Outcome responsibility 78
4.4 Conclusion 81

5 Climate harm and attribution science 84
5.1 Attribution methods: the probabilistic and the storyline approach 86
5.2 The reaction and criticism of the PEA community toward the storyline approach 89
5.3 The storyline approach and the criticism of overstatement 91
5.4 On how the probabilistic approach is affected by similar objections 94
5.5 Conclusion 99

6 Toward a rectificatory policy mechanism for loss and damage 109
6.1 Toward an adequacy-for-purpose view for attribution methods 110
6.2 An adequacy-for-purpose view for rectifying climate injustice 112
6.3 Distributing liability and achieving rectificatory justice 119
6.4 The political feasibility objection 122
6.4.1 Introducing and exploring the feasibility concern 123
6.4.2 Motivational and psychological aspects of the political feasibility objection 126
6.5 Conclusion 128

Final conclusions 133
Index 136

Acknowledgments

This book is the result of a long journey, which started with my PhD, matured during my postdoc phase, and finished during the first year of my appointment as Assistant Professor at the Erasmus University Rortterdam.

Various institutions have provided funding for the research time I spent in developing these ideas: La Caixa Foundation; the Doctoral Programme Climate Change – Uncertainties, Thresholds and Coping Strategies at the University of Graz; and the SNSF Project 'Principles of Ethical Decision-Making in Environmental Practice' at the University of Fribourg, the PhilETAS Research Group at the Karlsruhe Institute of Technology and the Erasmus School of Philosophy at the Erasmus University Rotterdam, where I am currently employed.

These institutions supported the Open Access fees for some of the papers on whose content I extensively rely for different chapters of this book. In Chapter 2, I drew upon A Minimal Capabilities-Based Account of Loss and Damage. *Journal of Human Development and Capabilities*, *25* (1), 170–193 (2024). In Chapter 3, I drew upon Backward-Looking Principles of Climate Justice: The Unjustified Move from the Polluter Pays Principle to the Beneficiary Pays Principle. *Res Publica* 29, 367–384 (2023). In Chapter 5, I drew upon my co-author paper with Douglas Maraoun, Overstating the effects of anthropogenic climate change? A critical assessment of attribution methods in climate science. *Euro Jnl Phil Sci* 13, 17 (2023). I am thankful to Austrian, German, Swiss and Dutch taxpayers for making accessible and independent scientific work possible.

I have discussed the ideas of this book with many colleagues at the different institutions I have worked for and at the different venues where I presented these ideas. All of them have been generous enough to devote their time and energies to improving my work. They know who they are, so I will not run the risk of forgetting any of them here. However, three people deserve especial attention. First, Lukas Meyer, for having built in Graz the academic community I always dreamt to be part of. Second, Clare Heyward, for her inexplicable trust in my work and for being the mentor I needed. Third, Eike Düvel, for our endless conversations about philosophy and his unconditional love.

My parents, Carmen Portela and Juan Carlos García taught me two things that have been essential for writing this, my first book. First, that those who are oppressed should not settle for charity when they have a historical justice claim; and second, that being resilient is the only option for those who cannot afford to fail. The first learning point summarizes the spirit of this book. The second one enabled me to make it through the ups and downs of writing a cohesive philosophical piece and pursuing an academic career. For those reasons, this book is dedicated to them.

Abbreviations

ACC	Anthropogenic Climate Change
AOSIS	Alliance of Small Island States
AR1	First Assessment Report of the IPCC
AR4	Fourth Assessment Report of the IPCC
AR5	Fifth Assessment Report of the IPCC
BPP	Beneficiary Pays Principle
COP	Conference of the Parties
EU	European Union
EWEs	Extreme Weather Events
GHGs	Greenhouse Gases
GMST	Global Mean Surface Temperature
IPCC	Intergovernmental Panel on Climate Change
LDC	Least Developed Countries
L&D	Loss and Damage (measures)
NIP	Non-Identity Problem
PPP	Polluter Pays Principle
SI(D)S	Small Islands (Developing) States
UN	United Nations
UNFCCC	United Nations Framework Convention on Climate Change
WIM	Warsaw International Mechanism

1 Introduction

The Day after Tomorrow, a science fiction movie released in 2004, depicts some of the effects of a series of extreme weather events that are impacting the planet. The northern hemisphere experiences hurricanes, tornados, rising sea levels, and extreme cold spells. The sudden destabilization of the climate leads to a new Ice Age where the world population will need to adapt to temperatures under −100°C. In our world, a cooling wave is not expected in the medium term. However, for some people in our world, it is already the day after tomorrow as the effects of climate change have come *lapping at their doors*.[1]

The crisis envisioned in *The Day after Tomorrow* is already made manifest in multiple real-world events, which are being triggered by the polar opposite cause: the global warming of our planet. Extreme weather events (EWEs) are increasing across the globe and are having significant negative effects on people's lives. In August 2016, heavy rainfall resulted in massive flooding in Louisiana, leaving thousands of buildings underwater.[2] Thousands of people faced severe health-related problems and 70,000 died in the 2003 European heatwave.[3] Ten years later, seven people died in Argentina due to the worst heatwave reported in the country until that date.[4] Slow-onset climate events also have (and are projected to continue to have) devastating consequences for people's lives. Communities living in polar regions or the Pacific Islands have already been affected by melting ice[5] or by rising sea levels.[6] In some cases, their habitats have been severely damaged, and, in other cases, their countries are facing a high risk of literally going underwater. Millions of people face hunger in Africa because of severe droughts caused by record high temperatures.[7] Climate change also contributed to the extreme floods in central Europe in summer 2021, which caused multiple fatalities, destroyed critical infrastructure, and left entire towns inundated with water and thousands of households without power and water. Heat deaths and a large increase in hospitalizations due to heat-related illnesses in July 2023 in the United States, China, and Europe have also been attributed to climate change.[8] Further, some studies suggest that the effects of droughts and low harvest have also triggered some of the fiercest political conflicts happening in the Middle East, such as the Syrian civil war in 2011 (Gleick 2014). As I write, residents of Lahaina, Hawaii, are reeling from the aftermath of the wildfire that devastated their town, with a current death toll of 110, which is expected to double. Simultaneously, Canada is experiencing the

DOI: 10.4324/9781003399889-1

worst wildfire season on record, with by far the largest area ever burned and fire activity expected to continue for several more weeks.[9] As I argue in this book, all these events are of utmost ethical concern because they infringe the human rights of many people on our planet.

For a long while, the negative effects of climate change were projected to materialize in the distant future. Talking about climate justice meant talking about justice toward future generations. Policymakers, political theorists, and philosophers focused on mitigation and adaptation policies to prevent the negative effects of climate change. However, the effects of climate change that are already occurring raise questions of justice in the here and now. These effects have been called 'loss and damage'. The duties or policies associated with addressing loss and damage have been termed L&D.[10]

Those who have suffer and are more likely to suffer the negative effects of climate change have been pushing for the recognition of historical responsibility for climate change (Friman and Hjerpe 2015; Calliari 2018). Historical responsibility for climate change has also had some pull in other areas of the policy and civic domain. Famously, Obama stated at the COP21 in Paris in 2015: 'I've come here personally, as the leader of the world's largest economy and *the second-largest emitter*, to say that the United States of America not only *recognizes our role in creating this problem*, we embrace our responsibility to do something about it'.[11] Obama's words suggest that how climate change came about should be a relevant consideration for the distribution of burdens in addressing climate change. This is, as I will call it, a rectificatory justice intuition.

Based on these considerations, this book has three main goals. First, it explains what should fall under the category of loss and damage, at least in a minimal sense. Second, it develops a rectificatory justice account for reparations for loss and damage. Third, it elaborates on how climate science can help in the design of a policy mechanism for addressing loss and damage. More generally, these three aims converge into the more practical goal of justifying and shaping a policymaking mechanism to address loss and damage.

The account I present in this work has two main theoretical commitments. First, it relies on the capabilities approach and a human rights framework to develop a minimal account for loss and damage. In a nutshell, I argue that loss and damage occur when climate change pushes people's capabilities below a sufficiency threshold, thereby infringing their human rights. Second, my account has a somewhat Nozickean inspiration. As Nozick did, I believe that 'sitting down at this late stage in history to dream up a description of the perfect society is not, of course, the same as starting from scratch' (1974, 313). This book starts from the assumption that there is something such as rectificatory justice, a form of justice that aims to repair some negative state of affairs by providing some normative force to the historical origins of that state. This intuition resonates with developing countries claims of historical responsibility as well as with Obama's words. This book argues that the anthropogenic and historical roots of climate change should play a role in the way we think about and frame climate justice. Doing so requires thinking about climate justice in terms of rectificatory justice. In this vein, my account of justice for loss and damage is grounded in a principle based on historical responsibility, namely, the Polluter Pays Principle (PPP).

This general introduction presents the context and some methodological features of this work. First, I explain the history of the emergence of L&D as a distinctive policy response, how this book contributes to developing an international policy mechanism for loss and damage, and the contested role of 'compensation' in climate negotiations. Second, I make some important terminological clarifications concerning the justice-related terms offered in the literature as well as those that I use in this book. Third, I explain the philosophical methodology behind this book. Fourth, I introduce two important assumptions concerning the scope of justice and background theories of justice and highlight three limitations of this work. Finally, the fifth section gives an overview of the general argument and a brief summary of each chapter.

1.1 An international mechanism for loss and damage

The adverse effects of climate change and the associated risks have been a matter of public debate since the creation of the IPCC in 1988. The aim of averting the harmful impacts of climate change has been at the core of the international response to this global crisis since 1992, with the adoption of the United Nations Framework Convention on Climate Change (UNFCCC). To avoid those impacts, mitigation and adaptation strategies have been put forward. Climate mitigation policies aim at preventing climate change, for example, by cutting GHGs emissions or by enhancing carbon sinks. Adaptation measures seek to moderate projected harmful impacts by adjusting the potentially affected ecological, social, or economic systems to actual or expected climatic stimuli and their effects or impacts. The harmful impacts of climate change that are already occurring, however, have given rise to L&D policies.

In this section, I present a brief history of loss and damage as a response to climate change before explaining how this book contributes to advancing our understanding and governance of loss and damage. Finally, I discuss the role of historical responsibility claims in the negotiations on loss and damage.

1.1.1 A brief history of loss and damage

Despite being on the table since the very beginning of UNFCCC negotiations, loss and damage issues have only gained traction in the last decade. The origins of loss and damage initiatives can be traced back to a proposal made by the Alliance of Small Island States (AOSIS) in 1991 in the negotiations during the run-up to the UNFCCC.[12] The concern of AOSIS at that time was the impacts of sea-level rise. AOSIS proposed to create an international insurance pool for compensating the most vulnerable small island and low-lying coastal developing countries for loss and damage arising from sea-level rise. Half of the pool was to be contributed according to each party's contribution to global emissions and the other half by their relative gross national product in the year prior to the year of contribution. This first proposal, however, was not accepted. In the first Conference of the Parties

(COP) held in Berlin in 1995, all that was achieved was an insurance mechanism to facilitate adequate adaptation for vulnerable countries. Consequently, the demands of vulnerable countries were subsumed under adaptation mechanisms.

After decades of the UNFCCC negotiations being focused on mitigation and adaptation, the topic of loss and damage gained currency in the wake of the Fourth Assessment Report (AR4) in 2007. This report highlighted the need to not only adapt to future impacts of global warming but also address those impacts that were 'already underway' (Parry et al. 2007). This warning gave strength to the 'Bali Action Plan' presented in 2007 at COP13 in Bali. The plan included the demand to not just address 'disaster risk reduction strategies' but also provide the means to address loss and damage (UNFCCC 2007). However, at that time, issues of loss and damage were still included as part of the adaptation strategy.

In 2008, AOSIS made a new proposal at COP14 in Poznan to create a loss and damage mechanism within the UNFCCC. This mechanism consists of three elements: insurance for the risks of EWEs; a rehabilitation and compensation mechanism to address slow-onset impacts; and a risk management component. However, this proposal was met with opposition from developed countries (especially the EU, Canada, and the USA), who argued against introducing any new institutions within the UNFCCC and opposed any mention of compensatory measures. This divide between developed and developing countries regarding the term 'compensation' has remained up to the present day.

It was not until 2010, at COP16, that the parties agreed on establishing a two-year work program to develop approaches to address loss and damage associated with climate change impacts in countries that are particularly vulnerable to the adverse effects of climate change (UNFCCC 2011 Decision 1/CP.16, 25–29). This program would assess the risks and approaches associated with loss and damage and clarify the Convention's role in implementing a solution. The goal of this initiative was to provide recommendations for COP18 in 2012.

After a series of meetings and campaigns championed by the Least Developed Countries (LDC) and supported by several NGOs (such as WWF, Oxfam, ActionAid, etc.), in 2012 at COP18 in Doha, it was decided to establish an international mechanism for loss and damage. This decision was a significant turning point, as it provided loss and damage issues with status as a component of the global agreements and institutions for tackling climate change. The final text, called 'Doha Gateway', did not frame future actions as 'compensation' but instead used language of 'rehabilitation'. It was not until COP19, which took place in Warsaw in 2013, that the parties established a concrete mechanism known as the Warsaw International Mechanism for Loss and Damage associated with Climate Change Impacts (known as the Loss and Damage Mechanism or Warsaw International Mechanism [WIM]). There, the parties acknowledged that loss and damage 'includes and, in some cases, involves more than that which can be reduced by adaptation' (UNFCCC 2014a). Although loss and damage issues remain within the framework of adaptation, some have interpreted that statement as a first step toward an eventual recognition of loss and damage as the third pillar for global climate policies (Kreienkamp and Vanhala 2017).

Today, the WIM is the central institution under the UNFCCC to 'address loss and damage associated with impacts of climate change, including extreme events and slow onset events, in developing countries that are particularly vulnerable to the adverse effects of climate change' (UNFCCC website).[13] The UNFCCC (2014a) assigned three main functions to the WIM: (1) enhancing knowledge and understanding of loss and damage and how to address it; (2) strengthening dialogue, coherence, and synergies among relevant stakeholders; and (3) enhancing action and support, including finance, technology, and capacity-building. At COP20, held in Lima in 2014, nine action areas were established for the working program (UNFCCC 2014b): (1) particularly vulnerable countries, (2) risk management, (3) slow-onset events, (4) non-economic loss and damage, (5) resilience, (6) migration and displacement, (7) finance, (8) collaboration with other bodies, and (9) development of a five-year rolling work plan.

The upholding of this mechanism is considered to be one of the main contributions of the Paris Agreement in 2015.[14] After Paris, the WIM reflected, to some extent, the demands of the LDC to work toward L&D as a specific strategy in global climate policies. Although loss and damage issues still remain within the framework of adaptation, the result of the Paris Agreement legitimized the exploration of categorizing losses and damages as a separate strategy beyond mitigation and adaptation. One crucial step was the inclusion of a stand-alone article (Article 8) for loss and damage in the final agreement. However, the desire of the LDC and small island developing states (SIDS) to include compensation under L&D measures remains contested. Not only did the agreement not contemplate compensation for climate-related loss and damage, it explicitly stated that 'Article 8 of the [Paris] Agreement does not involve or provide a basis for any liability or compensation' (UNFCCC 2015).

The next major step forward in loss and damage was achieved at COP27 in Doha in 2022, where the parties agreed to a loss and damage fund for vulnerable countries. To that purpose, a Transitional Committee was established to handle the negotiations in the development of the Loss and Damage Fund. As previously mentioned, three aspects are key to these conversations: what will be financed (i.e., what will count as loss and damage and what measures are necessary to address them), who will be the donors to this fund (i.e., how will loss and damage duties be distributed), and how we will identify where loss and damage occur. Even though the COP28 witnessed the proposal of a fund through voluntary start-up pledges from various nations, there is still a lot of work to be done to achieve a loss and damage fund that reflect that demands of climate justice.

In the next section, I explain how the results of this book contribute to understanding and grounding some of the features of an international mechanism for loss and damage and the possible fund associated with it. I identify five features for an international mechanism for loss and damage that map onto the functions of the WIM outlined earlier. In Section 1.1.3, I come back to the role of compensation and liability and argue that my account can also help in understanding this mechanism as a tool for rectifying the injustice associated with loss and damage as a matter of historical responsibility.

1.1.2 Contributing to an international mechanism for loss and damage

As mentioned in the previous section, the UNFCCC assigned three main functions to the WIM: (1) enhancing knowledge and understanding of loss and damage and how to address it; (2) strengthening dialogue, coherence, and synergies among relevant stakeholders; and (3) enhancing action and support, including finance, technology, and capacity-building.

This book contributes primarily to the first and third functions assigned to an international mechanism for loss and damage. In my view, in order to perform these two functions, such a mechanism should, at the least, do the following:

(a) define what counts as loss and damage and categorize its different forms;
(b) provide a variety of measures to address different forms of loss and damage;
(c) propose scientific tools to identify loss and damage due to climate change;
(d) propose principles of justice to assign duties to repair those losses and damages; and
(e) justify those operating principles.

While more will, of course, be said in the relevant chapters, I briefly summarize here how this work contributes to those tasks. First, this book addresses tasks (a) and (b) by defining a minimalistic understanding of what counts as loss and damage (i.e., as arising from human rights infringements). Further, it also embraces the common differentiation between economic and non-economic losses and damages and proposes some nuanced differences among the latter (Chapter 2). Second, this book addresses task (c) by delving into the science of attribution and analyzing the characteristics of each attribution method for purposes related to loss and damage as well as for providing the closest approximation of loss and damage due to climate change (Chapters 5 and 6). Third, it addresses task (d) by proposing to gather resources according to the PPP in light of the best available scientific information (Chapter 6). Fourth, it addresses task (e) by offering a justification for the PPP and responding to traditional objections to the principle (Chapter 4).

The work I engage in here supports the development of a mechanism for loss and damage based on historical responsibility. As we have seen, developing and vulnerable countries have pushed for such an understanding, while highly developed and historical emitter countries have rejected it. This book aims to explain why highly developed and historical emitters ought to agree to an international mechanism for loss and damage based on historical responsibility.

1.1.3 The contested role of compensation in climate negotiations

The use of the terms 'compensation' and 'liability' has been among the most troublesome issues in climate change negotiations. While this constitutes an essential issue for developing countries, it is considered a 'red line' for the developed world.[15] This issue has polarized the debate between developing and developed states. Developed countries, especially the United States, have firmly pushed back against framing issues of loss and damage in terms of 'compensation'. On the other side, there is a strong claim coming from developing countries for a new policy-making branch of loss and damage based on the idea of compensation for historical climate injustices (Friman and Hjerpe 2015; Calliari 2018).[16]

In an international law context, 'compensation' usually refers to a form of reparation for historical injustices, alongside restitution, recognition, or acknowledgment (Ivison 2006). According to Article 34 of the International Law Commission's Draft Articles on State Responsibility,[17] compensation is one way to make reparation or rehabilitation. Following this article's text, the first step in the process of reparation for past wrongs must be to restore the original scenario that existed prior to the commission of the wrongful act, for instance, by returning what was taken. 'Compensation' appears at a second stage, to amend or offset the consequences of the harm that stems from the past injustice, assuming that restoring what was taken is impossible. Finally, under this scheme, when neither of those options is possible, recognition or acknowledgment of both the injustice and the basic humanity and subjectivity of the victims is due.

In my view, the rejection of 'compensation' in climate negotiations has nothing to do with the specific meaning of the terms in International Law. That is, the term 'reparation' could just as well be rejected by those who reject 'compensation'. The disagreement is in fact much broader. Developed countries have rejected 'compensation' in climate change negotiations because it refers to liability for historical injustices. That is, by vetoing the terms 'compensation' and 'liability', developed countries seek to avoid any state responsibility and the subsequent obligation to bear the burdens of losses and damages based on their contributions to climate change.[18] What is at stake is the unwillingness of developed states to be held liable for their emissions, regardless of whether this takes the form of compensation or reparations. This book aims to explain that assigning historical responsibility to these countries is justified in the context of climate change.

In this book, I have avoided the term 'compensation' as far as possible because it can lead to some crucial misunderstandings. Instead, I use the terms 'rectification' and 'reparations' to refer to what should be the primary aim of L&D. Since the use of these terms and others might be ambiguous and often triggers misunderstandings, let me now clarify the meaning they have in this book.

1.2 A terminological note

In the literature on intergenerational justice, it is common to find terms such as compensation, restitution, rectification, or correction or their justice-specific variants compensatory justice, restorative justice, rectificatory justice, and corrective justice. Sometimes, these terms are used as synonyms. In other cases, they are used to express conceptual differences. In any case, their meanings and the differences among them are often not made explicit and clear. In this section, I explain how I will use these terms in this book in order to avoid any misunderstanding. With this, I do not mean to imply that this is the 'correct' meaning of these terms. Rather, this section aims to clarify the meaning of these words as they are used in the work I present here.

Let us start with 'compensation' and 'compensatory justice'. Compensation may be understood as a way of responding to the harm done to (or suffered by) someone. This response might aim to make this individual as well-off as he was before (*historical sense*), as well-off as he would have been (*counterfactual sense*), or as well-off as he should be according to some normative threshold (*normative threshold sense*). Unlike other responses to harmdoing (see below), compensation allows benefits and harms to be treated as reduceable to one single measure. According to this understanding, benefits can outweigh the harm, even if the harm itself remains. For instance, utilitarian theories make sense of this understanding of compensation by reducing both harms and benefits to one single measure of utility or happiness. If a person suffers an accident and loses a leg, compensating this individual means providing him with certain benefits that outweigh his harm. One can compensate this individual by giving him an orthopedic leg or a certain amount of money that would make him enjoy the same (historical or counterfactual) amount of utility despite the loss of his leg.

In climate justice literature, the term 'compensation' has been used to talk about responding to the suffering of harm, regardless of how that harm came about (Baatz 2013, 2016; Page and Heyward 2016). However, it has also been used to describe justice responses to culpable harmdoing (Meyer and Roser 2010; Meyer 2013). Further, claims of compensation have been situated within a distributive justice framework (Baatz 2013) and in opposition to distributive justice claims (Meyer and Roser 2010; Meyer 2013). For this reason, the use of the term 'compensation' or 'compensatory justice' may trigger some misunderstandings. This is reason enough to avoid these terms in this book.

A second reason to avoid framing loss and damage claims in terms of compensation is that compensation differs from other in-kind ways of responding to harm, such as restoration or restitution. For instance, restituting a person's health involves bringing their health to a state that can be defined in historical, counterfactual, or normative ways. Restorative justice could be understood as involving restitution in this sense. Note that, unlike compensation, restitution is a response that operates with the same kind of currency with which we frame the harm that an individual suffers. In this sense, an act of restitution would not involve simply giving money to the person who suffered harm as a response to that harm, but an act of compensation could involve such a measure. An unqualified use of the term compensation may suggest that my approach is indifferent to whether people are compensated in-kind or not. In order to avoid such misunderstandings, and because I believe that reparations should be in-kind as far as this is possible, I avoid the term compensation here.

Instead, I have chosen the term 'reparations' to refer to the kind of claims that fall under L&D duties and demands. Reparations can, in principle, be achieved through either restitution or compensation. However, in this book, I refer to demands of reparation that are closer to restitution than to compensation. In-kind responses to harm are given priority over other forms of responding to harm when considering what reparations require. I believe that reparations should be in-kind whenever possible and should be translated into other forms of compensation only when in-kind reparations are not possible. However, given that in-kind reparations might be impossible, compensation might be understood as part of reparations. For this reason, I have not chosen to frame the demands of justice that arise from climate change loss and damage in terms of restitution or restoration. Instead, I use reparations as an overarching concept that includes both responses to harm. That being said, my framework gives priority to in-kind forms of reparation.[19]

This book offers a framework for repairing losses and damages to rectify climate injustices. This raises a question about the difference between 'rectification' and 'reparation'. Thus far, I have explained the term 'reparation' as referring to a response to harmdoing whereby the duty-bearer has the duty to bring the victim to a certain state that is prescribed according to some historical, counterfactual, or normative standard, and according to the in-kind losses she has experienced. Let me now delve a bit more into rectification.

The term 'rectification' involves a response to an action carried out by an agent. Rectificatory justice is a kind of justice that requires acknowledging the sources of an injustice and the role of the actions leading to that injustice when deciding how to repair it. A rectificatory approach takes the actions that trigger the injustice as normatively relevant in remedying that injustice. A straightforward way in which a rectificatory approach acknowledges the sources of an injustice and the actions triggering is by targeting those *closely connected to the action* as the duty-bearers of rectification. The types of connections that I will explore in this work include *causing* the injustice and *benefiting* from it (Chapter 4). However, I do not disregard other possible forms of connection. The effects of events that do not qualify as actions (i.e., carried out by an agent) are not objects of rectification, although other agents can repair them.[20] The use of the term 'rectification' emphasizes the agent-focused dimensions of the approach I offer here and the relevance of the actions that are causing climate change.

To reiterate, a rectificatory duty is a duty to remedy or make up for the effects of unjust actions to which one is connected. Accordingly, a rectificatory claim is a claim to have the effects of an injustice repaired in a form that acknowledges the actions attached to the injustice. This requires that the claims of rectificatory justice are addressed to those closely connected to the injustice. I use the term 'rectificatory duties' and 'rectificatory claims' to include various types of duties and claims, ranging from symbolic reparations to material reparations.

The term 'corrective justice' has also been used as an overarching term in the literature on historical injustices (Shelby 2013; Butt 2017). Corrective justice covers the two sources of duties that are addressed in this book, namely, causing an injustice and benefiting from an injustice. However, this term is drawn from legal literature (Coleman 1995). In my view, this fact generates a certain level of confusion about whether the empirical and epistemological demands of corrective justice need to satisfy the legal standards in place in the policy domain. For this reason, I have decided to use a more neutral term, which pertains only to the moral and political debate.

To summarize, this work is concerned with the justification of reparations for loss and damage from climate change as a matter of rectificatory justice. That is, reparations are addressed here in a way that acknowledges the role of the actions triggering the injustice. Based on this, I justify the duty to provide both symbolic and material reparations for different economic and non-economic aspects of loss and damage. Since I understand the exclusion of the words 'compensation' and 'liability' in climate negotiations to be grounded in a rejection of backward-looking and rectificatory approaches to climate justice, this work disputes the rejection of those terms in the context of loss and damage.

1.3 Philosophical methodology

Having explained the aim, scope, and motivation of this book, I turn now to methodological considerations. In this section, I present both the general philosophical methodology adopted here and a methodology more specific to the area of climate justice.

1.3.1 General methodology: reflective equilibrium

Throughout this book, I employ the method of reflective equilibrium. This method was famously introduced by John Rawls (1971) and has been widely used in analytic political philosophy. Reflective equilibrium consists in working back and forth between our considered judgments about particular cases, together with certain moral principles and background theories, until we reach coherence among them. The aim of the method is to come up with a theory of justice whose internal elements cohere with each other.

Norman Daniels (1996) has specified and described the elements of reflective equilibrium as follows:

- Considered judgments are the kinds of judgments that individuals make on moral issues when they reflect upon them under conditions conducive to forming rational judgments.
- Moral principles are moral rules attached to different philosophical positions such as 'prioritize the well-being of the least well-off' or 'act as if the maxims of your action were to become through our will a universal law of nature'.
- Background theories are a set of propositions belonging to different knowledge areas that support both the considered judgments and the moral principles, for instance, by providing the empirical evidence required for those judgments or principles. Sociological, psychological, or physical theories can be included here.

When employing this method, these three elements are all revised in view of one another to achieve an acceptable coherence – or reflective equilibrium – among them all. For instance, we can think of the process as starting from considered judgments. One then compares those considered judgments with certain moral principles, taking background theories into consideration. If these elements do not cohere with each other, one may revise or reject some of them. Further, one may also add new beliefs that were not previously included.

Let me provide a rough example of how this work employs the method of reflective equilibrium. I begin this project with the considered judgment that those who cause climate change–related harm should bear the burdens of repairing it.[21] This considered judgment is supported by certain background theories, such as scientific theories that describe climate change as being caused by anthropogenic forces. I also revise this judgment according to the moral principle that people can only be held morally responsible if they act under conditions of culpability or blameworthiness. This moral principle is then compared with legal theories of strict liability, modified according to the considered judgment, and so forth.[22]

I am aware that attaining a perfect reflective equilibrium is a nearly impossible task, nor do I claim to do so in this work. However, this method informs the work done in this book to a significant degree, even if further elements could be considered to challenge or modify the results I present here.

1.3.2 Climate justice specific methodology

Methodologically, we can approach climate justice from an isolationist or an integrationist perspective. In this book, I adopt a 'moderate integrationist' approach.

Following Caney (2018), integrationism and isolationism differ in how they answer the following questions: should we include other considerations of justice (such as people's entitlements to food or health) in the design of a climate justice scheme? Or should we instead treat climate problems and responsibilities in isolation from other considerations of justice?

An isolationist approach answers the first question in the negative and the second question in the affirmative. It holds that we should apply principles of justice to a given good, in this case, climate-related goods – such as emission permits – in isolation from other considerations of justice. One clear example of isolationism is the equal per capita emissions approach to the distribution of global emissions, which states that every person has a right to the same level of GHG emissions. This approach is isolationist because GHG emissions are treated here as the only relevant good to be distributed, regardless of people's non-climate-related entitlements.

Integrationists or, more precisely, strong integrationists, hold that we should treat climate change–related considerations in conjunction with other justice-related considerations (such as a global difference principle or a commitment to basic rights). To see what a strong integrationist proposal may look like, we can consider an integrationist criticism of the isolationist equal per capita approach. For strong integrationists, what matters is not that every human being is able to emit the same amount of emissions, but rather that every individual is able to enjoy basic needs and rights associated with a decent human life. In this sense, emission permits should be distributed according to how they contribute to meeting people's basic needs. This method begins with the basic needs and rights that people have and then proposes how emission permits should be distributed. Accordingly, an integrationist approach might support an unequal distribution of emission permits, if such a distribution enables people's basic needs and rights to be satisfied.

We could differentiate between strong integrationism and isolationism in climate change policies in relation to the goals they aim to achieve, on the one hand, and, on the other, to their allocation of climate-related burdens for mitigation, adaptation, and loss and damage (Baatz 2016, 138). First, climate policies would be isolationist if they aim to rectify the effects of climate change on people's lives, and only those effects. They would be strong integrationist if they were to aim at securing an overall level of the currency of justice, regardless of whether doing so would also imply addressing other injustices, for instance, if they sought to secure human capabilities or basic goods, regardless of whether those were previously unsecured because of other injustices apart from climate change.

Second, climate policies can be categorized as isolationist or strong integrationist depending on whether they consider other issues of justice when allocating climate burdens. We can also apply isolationist or strong integrationist criteria for each type of climate policy. In this sense, mitigation policies, such as equal per capita emissions policies, are isolationist because they only take the production of GHG emissions into account when distributing emissions permits. For instance, we can imagine a country that has produced a low quantity of GHG emissions across time but whose inhabitants enjoy a high level of welfare. An isolationist approach would favor an equal distribution of GHG emissions among countries and thus allocate the same amount of GHG emission rights to this country as to any other. This is because what matters is the allocation of climate-related assets, namely, GHG emissions, regardless of other considerations of justice (for instance, the level of general welfare that the inhabitants of that country enjoy). Mitigation policies are integrationist if they allocate more GHG emissions to countries with lower welfare levels regardless of their historical GHG emissions, on the assumption that GHG emissions rights would help to increase their welfare.

Likewise, adaptation policies would be isolationist when it comes to allocating climate change–related burdens if they were to allocate adaptation payments exclusively according to individual contributions to climate change. For instance, those who have contributed the most to climate change would bear heavier climate change–related burdens than those who have contributed less, regardless of general distributive justice concerns. Adaptation policies would be strong integrationist if they were to distribute climate change–related burdens based on the general distribution of other goods, for instance, according to the capacity to develop and implement adaptation policies that benefit those most affected by climate change.

Simon Caney has left the door open for a third alternative, which he terms moderate integrationism. Moderate integrationism differs from strong integrationism in that it takes one principle to be more prominent than others while remaining open to other considerations of justice. The example used by Caney is far removed from considerations of climate justice, but it gives us an idea of what he has in mind. He writes:

> One might think that the appropriate principle for a speeding offence is something like 'Those who speed should be punished and the more that they

> exceed the limit the more severe the punishment'. Now suppose that someone is speeding, we might think that this is the relevant principle; but we might also add that it can be qualified by taking other normative considerations into account. For example, we might take into account whether they had a pressing medical need or if they are being chased by someone seeking to hurt them. There is nonetheless a recognizable principle of justice that applies here and, so to speak, plays the dominant role.
>
> (Caney 2018)

In my view, a moderate integrationist approach should take not only certain principles as dominant but also certain harms. For instance, in the case of climate change, a moderate integrationist approach would take climate change–related harms as dominant (see below). The approach I present here follows this methodology. I defend the PPP as the dominant guiding principle for climate justice, but I also consider other issues that qualify the application of that principle.

This moderate integrationism shares some similarities with both isolationism and strong integrationism. First, it shares with isolationism that climate justice policies aim to address the impacts caused by climate change, and, in principle, only those impacts.[23] This does not mean that this should be taken as the end state, that is, a place where other considerations of justice no longer apply. What I maintain is that these are the demands of *climate justice*. Further improvements in people's lives could well be justified if they do not meet a certain just threshold of well-being. But when we engage in the project of providing further aid, we enter into a different realm. We leave the realm of climate justice and enter into the realm of global distributive justice. Second, moderate integrationism shares with strong integrationism the idea that the application of climate justice should be qualified by other considerations of justice. That is, we should not require individuals to pay for their emissions to the point that they fall below a relevant threshold of justice, which I take to be sufficientarian.

Alongside these similarities, moderate integrationism also differs from both isolationism and strong integrationism. Unlike (at least certain readings of) isolationism, moderate integrationism claims that considerations of global justice should constrain considerations of climate justice, especially if the application of climate justice principles would cause the same type of harms that they are intended to alleviate. For instance, a principle of climate justice whose application would cause the duty-bearers to fall below a certain level of well-being ought to be rejected. Unlike strong integrationism, moderate integrationism claims that polluters have merited certain duties because of their contribution to climate injustices and that they have a duty to rectify those injustices that prevail over other considerations (for instance, the legitimate expectations they might have).

I am aware that the versions of isolationism and strong integrationism presented here might be too simplistic. Scholars endorsing one or the other approach might in fact have approaches that are similar to my moderate integrationism. Nonetheless, the contrast between my approach and the simplified description of those approaches is helpful to demonstrate the methodology used in this book.

1.4 Assumptions and limitations

This book operates with certain assumptions and limitations that influence the understanding of the arguments and claims offered here. Although it would be impossible to call attention to all of them, this section will explain major assumptions and their corresponding limitations. I make two significant assumptions. The first concerns the scope of justice. That is, who are the recipients and the duty-bearers of climate justice? I assume that the recipients of justice are individual people and that the relevant duty-bearers are states. Here, I explain these assumptions, with a special emphasis on how to understand states as the duty-bearers of climate justice. The second assumption concerns my embrace of sufficientarianism as a background theory of distributive justice and a sufficientarian understanding of the operative threshold notion of harm.

1.4.1 The scope of justice: recipients and duty-bearers of climate justice

The scope of justice addresses who the duty-bearers and the recipients of justice are. The approach I offer here combines the idea that individuals are the recipients of justice with the idea that states are the relevant duty-bearers. Taking individuals as the recipients of climate justice may be uncontroversial since moral individualism (the idea that individuals are the ultimate unit of concern) is widely accepted in Western liberal moral and political philosophy. This idea does not preclude the possibility that individuals have interests and claims of justice as members of certain groups. Yet, these claims are understood as ultimately grounded in their individual interests, not in the interests of the collective entities. Although I do not exclude the possibility of extending my analysis to collective entities, that discussion lies beyond the scope of this book.

Various climate scholars have also embraced the idea that states can be taken to be the relevant duty-bearers of climate justice (Zellentin 2015; Gardiner and Weisbach 2016; Page 2012; Heyd 2017; García-Portela 2019; Francis 2020). Usually, this is taken as a starting assumption. However, at a deeper level, it may appear as philosophically more controversial, especially when it comes to attributing rectificatory duties.[24] To assign responsibility for certain actions, we must be able to connect a particular outcome to a particular agent's actions, and some may argue that states do not qualify as the kind of agents that can be ascribed with agency in the relevant sense. Without this sense of agency, any attribution of responsibility for action is rendered invalid. Yet, everyday political discourse often involves judgments that hold states responsible for their contemporary and historical actions. In David Miller's words:

> We say that Russians are responsible for the civil war in Chechnya, Israelis for the fate of Palestinian refugees, and Americans for their excessive contribution to global warming. We also make judgements about events that have occurred in the national past: we hold Britons responsible for the deaths of one million Irish people in the potato famine, Turks for the Armenian genocide, and Germans for the Holocaust.
>
> (Miller 2007, 111)

In this section, the question I want to address is how we might understand the claim that states are the duty-bearers – which can be held responsible for both the contemporary and past actions – if we were to hold onto the intuitions behind these everyday political judgments. While I cannot provide an extensive explanation of this issue here, I outline different ways in which this claim can be understood. In doing so, I also point to the implications of these different understandings for some of the arguments I make in this book.

There are at least two complications with the claim that states are the relevant duty-bearers. The first complication is that states vary in their formal and institutional structure over time. Further, and more importantly, some states that perpetrated injustices in the past no longer exist (for instance, the Nazi state) or have been divided into different states (e.g., the USSR). This raises questions such as whether the historical responsibility of such states fades away with their institutional structure or whether states that have been newly formed start from scratch and have no historical responsibility.

We can address this concern by spelling out how states are taken to be the relevant agents. This claim can be understood instrumentally rather than metaphysically. That is, instead of understanding the state as a metaphysical unit that continues over time, the state could be taken to be the institutional unit that discharges the duties of a more fundamental entity. A common and straightforward approach is to consider the state as acting in the name of a nation. Roughly described, a nation 'is a community of people who share an identity and a public culture, who recognize special obligations to one another and value their continued association, and who aspire to be politically self-determining' (Miller 2004, 243, similar in Miller 2007).

In my view, nations thus described can be understood as transgenerational entities since identities and public culture can be shared among non-contemporaries. Moreover, people living in the present recognize special obligations toward future members of the same nation, as well as toward past members (for instance, obligations to praise the deeds of past generations). Furthermore, people value and work for the continued existence of their nations and strive toward self-determination or to maintain their self-determined nature in the future. Understood in this way, a nation is a more fundamental entity that may survive changes in the state's institutional and administrative structure,[25] that may exist without a state, and that may emerge as a new state within a previously existing state (Anderson 2006). In addition, a state may exercise authority over several different nations and may have to discharge the duties of various nations.

The second complication was mentioned earlier: it is unclear whether states are able to be the kind of agents that can be held responsible or accountable for their actions, even if they are understood as representing nations. The claim here is that it is unclear whether a collective and transgenerational entity, such as a nation, has the necessary features to be considered an agent to whom we can attribute certain duties for past actions. Since this is possible for individual people, the question is whether we can hold an analogy between individual people and collective entities of this sort in applying principles of historical responsibility.

One way to address this concern is by offering an account that makes sense of the agency of such a collective transgenerational entity. I shall term this strategy the One Entity Strategy. In a nutshell, according to this argumentative strategy, there is just one entity (a nation) that has persisted over time and bears the responsibility for certain actions that lead to rectificatory claims. There is a vast literature on collective responsibility that addresses this question (Smiley 2017). Some proposals brought forward in this field may be able to provide an account of how a national transgenerational entity can meet the conditions of agency necessary to attribute responsibility and rectificatory duties.[26] Unfortunately, space does not allow for a full discussion here of how such an account would work. What can be said is that if such an account is successful, our everyday judgments on state responsibility can be explained according to the One Entity Strategy as follows: A state is an institutional structure representing a nation that may have a long history. That nation may be responsible for some past actions such that it bears rectificatory duties. The state that represents that nation today is obligated to discharge those duties as representative of that collective and transgenerational entity. This is one way of understanding the claim that states are the duty-bearers of justice.

However, in the absence of a concrete account of transgenerational and collective responsibility applied to nations, we might want to explore other possibilities. A different strategy is to consider two groups of people, one of past individuals and another of current individuals, both belonging to the same transgenerational national community. The claim, then, is that people living now are obligated to discharge the duties associated with past community members' responsibility for their past actions. Thus understood, contemporary individuals' duties are based on some form of vicarious responsibility for what past members of their communities did. According to this approach, the state is the duty-bearer in the sense that it is the institutional unit responsible for discharging the duties of its current citizens. I shall term this account the Two Groups Strategy.[27] Such an account requires that we explain why belonging to the same nation or state gives rise to duties associated with the actions undertaken by past individuals. I explain here two different ways to answer that question, one proposed by David Miller (2004, 2007) and another offered by Janna Thompson (2002, 2017).[28]

David Miller (2004, 2007) has offered different explanations for why belonging to the same nation gives rise to duties associated with past actions, depending on the kind of rectificatory duties at stake. Here, I focus on those associated with having caused an injustice that requires both material and symbolic reparations. According to Miller, individuals alive today and belonging to a certain nation may have material rectificatory duties related to injustices perpetrated by past members of the same nation because they also enjoy goods associated with those past members' actions. In Miller's words, the rationale here is that 'present-day As [members of nation A]. . . cannot disown the policies in question [associated with injustices] while continuing to "own" those other policies and practices of older generations of As that now provide them with advantages' (Miller 2007, 155). This argument is one of consistency: the enjoyment of the advantages provided by previous generations leads to a duty to also accept their liabilities since one cannot coherently accept the former while rejecting the latter.

This explanation can be extended to symbolic reparations. As Miller argues, demands of apologies seem to presuppose ties of identification between generations, and not just the inheritance of certain goods. However, we can also use an argument of consistency here. Following Miller, 'one cannot, morally speaking, identify with the positive past achievements of one's nation and take pride in them without at the same time acknowledging responsibility, and the need to apologize, for past actions that were harmful to others' (Miller 2007, 158–59). Moreover, 'since national pride is a widespread phenomenon, so too is the potential scope of national apology' (idem). Note that, under this interpretation, those apologies are (partly) vicarious apologies.

In a nutshell, according to Miller's explanation, one cannot consistently accept the advantages of belonging to a certain nation without accepting the liabilities, and one cannot take pride in the deeds of one's forebearers without also needing to apologize for them. To summarize, states are the duty-bearers of intergenerational justice in the sense that they are the institutional units responsible for discharging the duties of their current citizens related to the liabilities and symbolic duties merited by their nationals in the past.

Janna Thompson has provided a different account (Thompson 2002, 2017, 2006). Thompson starts with the assumption that citizens of a state want or should want their state to be just in the present and in the future

> for the sake of their children and grandchildren or the groups with whom they identify, or because they value maintaining respectful relations with other intergenerational communities, or simply because they value the perpetuation through the generations of just institutions and practices.
>
> (Thompson 2017, 57)

This concern should cause them to demand that their successors both maintain just institutions and repair the injustices they might commit. Following a principle of 'like cases treated alike', she argues that citizens of that state should also accept similar burdens concerning their state's past deeds. If they value justice in the present, want it to be held in the future, and want to ensure that future generations repair their acts of injustice, they should also rectify the injustices committed by their predecessors. Alongside their duty to promote and maintain just institutions in the present, they also have a duty to rectify their communities' past injustices.

Note that, unlike Miller's account, Thompson's account does not require that current citizens of a state enjoy certain advantages from living in that state, nor that they feel pride in past generations' deeds. All that is required is that they are committed to justice in their society, which they have reasons to want. Summarizing Thompson's approach, states are the duty-bearers of intergenerational justice that can be held accountable for past (and present) actions in the sense that they represent a group of present-day individuals who exhibit a commitment to justice that concerns the present, the future, and the past.

These two strategies (the One Entity Strategy and the Two Groups Strategy) can be used to elaborate the claim that states are the duty-bearers of climate justice. Note that climate change is a phenomenon caused by both past and present emissions. First, according to the One Entity Strategy, states are the relevant duty-bearers of climate justice in the sense that they are the institutional units that are obligated to discharge the rectificatory duties of the nation(s) they represent, which is conceived as a transgenerational and collective entity. These rectificatory duties emerge from this collective entity's contribution to climate injustices through historical and current emissions. Second, according to the Two Groups Strategy, states are the relevant duty-bearers of climate justice in the sense that they are the institutional units that are obligated to discharge the rectificatory duties of their citizens merited by both their own emissions and the historical emissions of past generations. In the second case, present-day people are obligated to discharge, through their states, past generations' duties related to their emissions in order to be consistent with their acceptance of the goods acquired from those past generations and with the pride they feel toward them (following Miller's account), or to be consistent with their commitment to justice in the present (following Thompson's account).[29] Note that these two strategies do not need to be invoked for the emissions of people who are alive today.

This book does not commit itself to one of the above strategies over the other. Instead, I assume that the claim that states are the duty-bearers of climate justice can be interpreted according to either of them. However, I would like to close this section with an explanation of the implications of using one approach or the other. In Chapter 2, the book argues that states have duties of symbolic reparation for the harms caused by their emissions. This claim has different interpretations depending on the strategy one adopts. According to the One Entity Strategy, the claim requires the clarification that the nations represented by those states have duties of symbolic reparations. According to the Two Groups Strategy, the claim requires the clarification that the individuals living in those states have duties of symbolic reparations. Thus the claim that the states have duties of symbolic reparations (such as apologies) is simply a statement that should be interpreted according to either of the strategies I have laid out here.

Nevertheless, adopting one or the other strategy has different implications. Suppose we adopt the One Entity Strategy. In this case, we must assume that nations (understood as collective transgenerational entities) are the kind of entities that not only have agency but can also apologize. Suppose we adopt the Two Groups Strategy. In this case, we must assume that symbolic reparations and, more specifically, apologies are (partly) vicarious because individuals alive today are, at least partly, apologizing for the actions of others.[30] This also assumes that agent-regret, which will be considered as the trigger for duties to apologize, can be vicarious as well. None of these implications are clarified any further in this work, and, for that reason, they operate as underlying assumptions.

Moreover, in this book I argue that polluters have broader rectificatory duties that concern the reparation of the harm caused to victims of loss and damage. If we follow the One Entity Strategy, we should read my conclusions as implying that 'polluters' here refer primarily to nations, conceived as collective and transgenerational entities. In this sense, states have rectificatory justice duties because they represent that nation. As before, given the argument I offer in that chapter, this requires the assumption that this transgenerational and collective entity is able to act according to reasons. If we follow the Two Groups Strategy, we should read my conclusions as implying that present-day individuals bear rectificatory justice duties for their own polluting actions and for past polluters' actions. With regard to past emissions, technically speaking, past polluters are the ones who initially bear rectificatory justice duties. Present-day individuals inherit those duties for the reasons provided by the Two Groups Strategy. In this case, states are the relevant duty bearers in the sense that they are obligated to discharge the duties of their current citizens for their own actions as well as for past individuals' polluting actions.

Throughout this book, I use the term 'state' to refer to the duty-bearers of climate justice. However, the use of this term shall be properly understood according to any of the strategies that I have outlined in this section. That is, states can be understood to be the duty-bearers of climate justice as representatives of an overarching collective and transgenerational entity, namely, a nation. Alternatively, states can be understood to be the duty-bearers of climate justice as representatives of a community of people alive today who have certain duties associated with the past actions of other people, to whom they relate in certain ways. Note that, strictly speaking, in neither of these interpretations are states the primary and fundamental moral unit. However, they can be understood as the political institutions responsible for discharging the duties of a more fundamental moral unit, however this may be understood. I have chosen to use the term 'states' to refer to the relevant political duty-bearers because it better accommodates these two interpretations. I leave up to the reader whether to understand states as the duty-bearers of climate justice according to the One Entity or the Two Groups strategy.

1.4.2 Sufficientarianism: background theory of distributive justice

The framework I propose in this book operates with sufficientarianism as a background theory of distributive justice. I also embrace a sufficientarian interpretation of the threshold of harm grounding rectificatory justice claims. In this section, I explain sufficientarianism as a theory of distributive justice, how such a theory can also ground a threshold of harm, and how both considerations will be operating throughout this book.

Sufficientarianism claims that what is important in terms of justice is whether people have enough to live decent lives (Frankfurt 1987). Unlike other theories of distributive justice, the most relevant normative consideration is *not relational*. According to sufficientarianism, what is relevant from a distributive justice point of view is not whether a person has *more or less than others*, but rather whether that person has enough in terms of some objective currency, a sufficient level of which is required to live a decent life. The delimitation between 'having enough' and 'not having enough' is usually expressed in terms of a sufficientarian threshold. Sufficientarianism can be understood as embracing a positive thesis and/or a negative thesis. According to the positive thesis, distributive justice requires that people have enough to live a decent life. According to the negative thesis, distributions above the threshold are not morally relevant or, at least, are much less relevant (Casal 2007). In the following, I explain how to understand these claims in this book.

This book endorses the positive sufficientarian thesis as a background theory of distributive justice. This background theory limits the demands of rectificatory justice defended here. To be more specific, according to my account, claims of rectificatory justice do not outweigh sufficientarian distributive justice considerations. This means that those obliged to rectify climate injustice should not fall below a threshold of sufficiency as a consequence of discharging their rectificatory justice duties.[31] This latter point is also implied by the negative thesis since I argue that distributive considerations that fall above the threshold might be outweighed by other considerations, such as the duty to satisfy the reasons one has left unsatisfied (see Chapter 4).

As I said, I also embrace the negative sufficientarian thesis, according to which distributions that fall above the threshold are not morally relevant or, at least, are much less relevant. I believe that, for the purposes of this book, one need not embrace the strongest interpretation of the negative thesis (i.e., the idea that distributions above the threshold are not morally relevant at all). It is enough to acknowledge that distributions above the threshold are much less relevant than distributions below the threshold. For this book, the relevant implication of this negative thesis is that above the threshold (but not below) other considerations of justice may override the relevance of further distributive justice considerations that may apply above the threshold (such as egalitarian or prioritarian considerations of justice). Throughout this book, I argue that considerations of distributive justice that fall above a sufficientarian threshold level cannot override grounds for rectification.[32]

Moreover, in my view, even if we attribute greater moral importance to distributions above the threshold than those implied in this book, this will not change the conclusion that most historically high-emitting countries should discharge their rectificatory climate justice duties.[33] Given their high levels of welfare above the threshold, discharging their rectificatory duties would rarely bring about an unfair above-threshold distribution. These considerations are contingent on the current global distribution of resources and may change over time.[34] However, for the time being, we need not be concerned that my results might bring about highly unjust distributions above the threshold.

Note that the approach developed here does not fully satisfy sufficientarian distributive justice demands because I argue that those causing climate change should *exclusively* rectify the (below-threshold) harm they have caused. However, people who suffer from climate change often already find themselves below a threshold of sufficiency before climate change–related impacts materialize. Exclusively rectifying the impacts of climate change may still leave them below the relevant threshold. In some circumstances, rectifying climate injustice might bring people to a threshold of sufficiency, but only contingently so. In other circumstances, rectifying climate change–related harm would still leave people below a threshold of sufficiency. In light of this, I believe that my approach could be complemented with an overarching sufficientarian theory of global distributive justice that allows other justice demands to be satisfied.[35]

1.5 Summary of chapters

This book has five chapters, through which I develop different relevant aspects for advancing a rectificatory justice account for loss and damage. Let me provide a brief summary of each of them.

Chapter 2 develops an account of what constitutes, at least minimally, loss and damage. It argues that loss and damage occur, *at least*, when climate change disrupts people's lives by pushing them below a sufficient standard in their opportunity to enjoy the central aspects of a dignified, flourishing life (or capabilities) and that such a disruption constitutes an infringement of their human rights. I call this view the minimal capability-based account of loss and damage, which goes hand in hand with an account of loss and damage based on human rights. The chapter also situates loss and damage in comparison to other climate change measures and provides a categorization of each type of loss and damage (from economic to non-economic loss and damage). Finally, the chapter proposes reparative measures to address different manifestations of loss and damage.

Chapter 3 explores two principles of rectificatory climate justice: the Polluter Pays Principle (PPP) and the Beneficiary Pays Principle (BPP). According to the first one, rectificatory justice duties should be distributed according to historical responsibility tracked by each nation emissions records. According to the second one, rectificatory justice duties should be distributed proportionally to the benefits acquired through emissions. This second principle was proposed in the literature to circumvent the problems faced by the first principle, namely, the Causation Objection and the Excusable Ignorance Objection. In this chapter, I argue that we do not have good reasons for rejecting the PPP and embrace the BPP based on those objections because this second principle is also subjected to the same concerns giving raise to those objections. Thus, a rectificatory justice account could be based on any of those principles, as long as it is able to provide a convincing solution for those objections.

Chapter 4 engages with the task of providing a justification for the PPP that can circumvent the main objections pressed against this principle, thereby providing the grounds for a rectificatory climate justice account. This justification is provided by what I call the Continuity Account. In a nutshell, the Continuity Account claims that polluters should bear the duties of addressing loss and damage because those duties stem from a previously unsatisfied duty of not infringing human rights. Also, this chapter explains not only how this account can circumvent the main objections against this principle but also how it improves existing accounts to distribute climate change–related duties based on a direct principle of historical responsibility.

Chapter 5 delves into the problem of how to identify climate harm. That is, it presents and discusses the main attribution methods to link environmental loss and damage with climate change: the probabilistic approach and the storyline approach. It engages with the worry that the storyline approach overstates the effects of anthropogenic climate change (ACC), given the important implications this fact would have in liability contexts. However, this chapter shows that this worry is unjustified and that there are no reasons to dismiss the storyline approach and favor the alternative account based on this consideration.

Chapter 6 completes my account of rectificatory climate justice on two remaining fronts: it explains how to evaluate and decide on attribution methods based on considerations other than possible overstatements and it provides an account of how liability could be distributed within a policy mechanism for loss and damage. First, this chapter provides an adequacy-for-purpose argument for choosing between attribution methods in a way that assigns priority to the probabilistic approach for its affinities with legal reasoning in liability contexts. Although this argument is non-conclusive, I argue that it provides some relevant reasons to prefer one approach, namely, the probabilistic approach, over the alternative. Second, this chapter explains how a policy mechanism for loss and damage could function following the input provided by attribution studies and how liability could be distributed once the results from attribution methods are considered. Finally, the chapter addresses the problem of political feasibility of a historical responsibility–based account and develops a somewhat speculative argument to counter this objection. With these insights, the chapter brings the development of a policy mechanism for rectificatory climate justice for loss and damage one step further.

Notes

1 The expression 'lapping at their doors' recalls the unfortunate words of the Australian minister of immigration, Peter Dutton, in 2015. After Tony Abbott, then Australian prime minister, complained that the Pacific Islands Forum in Papua New Guinea was running late, Dutton said: 'Time does not mean anything when you are, you know, about to have the water lapping at your door'. Many residents in Kiribati found Dutton's words outrageous. For further details and information, see: https://www.theguardian.com/australia-news/2015/sep/11/peter-dutton-jokes-with-tony-abbott-about-rising-sea-levels-in-pacific-nations; and https://www.theguardian.com/world/2017/oct/23/waiting-for-the-tide-to-turn-kiribatis-fight-for-survival.

2 For detailed information, see: https://www.bbc.com/news/world-us-canada-37121404.

3 For detailed information, see: https://www.carbonbrief.org/study-links-heatwave-deaths-london-paris-climate-change.

4 For more detailed information, see: https://www.independent.co.uk/news/world/americas/seven-people-die-in-the-worst-heat-wave-ever-recorded-in-argentina-9032202.html.

5 One of the most popular examples of this sort is the Inuit community, who in 2005 filed a petition seeking compensation for the violation of their human rights due to climate change.

6 An example of this is the low-lying small island of Kiribati. For more detailed information, see: http://www.climatehotmap.org/global-warming-locations/republic-of-kiribati.html.

7 For more detailed information, see: https://www.theguardian.com/environment/2016/mar/16/drought-high-temperatures-el-nino-36m-people-africa-hunger.

8 For detailed information, see: https://www.worldweatherattribution.org/extreme-heat-in-north-america-europe-and-china-in-july-2023-made-much-more-likely-by-climate-change/.

9 For more detailed information, see: https://www.cbc.ca/news/politics/canada-wildfire-season-worst-ever-more-to-come-1.6934284
10 Here, I follow the convention of capitalizing the policy term 'L&D' while using lowercase 'loss and damage' (or, the plural, losses and damages) for the impacts themselves (Mechler et al. 2019). When referring to both at the same time (for instance, to the topic itself), I use the singular in lowercase 'loss and damage'.
11 Emphasis mine. For more information, see: https://www.forbes.com/sites/jeffmcmahon/2015/11/30/obama-says-u-s-accepts-its-responsibility-for-climate-change/?sh=173efb711fe5.
12 AOSIS Proposal to the AWG-LCA: Multi-Window Mechanism to Address Loss and Damage from Climate Change Impacts. Alliance of Small Island States. Available at: http://unfccc.int/files/kyoto_protocol/application/pdf/aosisinsurance061208.pdf
13 See https://unfccc.int/topics/adaptation-and-resilience/workstreams/loss-and-damage/warsaw-international-mechanism.
14 The second main pillar of the Paris Agreement might be regarded as the commitment to limiting the global temperature rise to below 2°C and the intention to work toward the achievement of a 1.5°C limitation.
15 For instance, before COP21, the US Secretary of State, John Kerry, declared that framing loss and damage as an issue of compensation would 'kill the deal' (see https://www.rollingstone.com/culture/culture-news/john-kerry-on-climate-change-the-fight-of-our-time-50220/).
16 Dr. Saleemul Huq became an almost historical figure among activists and researchers from developed countries for his work on loss and damage and his insistence that polluting nations should pay for compensation, both in scientific articles and in the media. Some of the outreach pieces reflecting these ideas include: https://www.icccad.net/dr-saleemul-huq-media/its-time-to-make-polluters-pay-for-climate-damages/ and https://www.icccad.net/daily-star-articles/make-polluters-pay-for-loss-and-damage-from-climate-change/.
17 International Law Commission Articles on Responsibility of States for Internationally Wrongful Acts, GA Res 56/83, UN GAOR, 6th Comm, 56th session, agenda item 162, UN doc A/REST/56/83 art. 34.
18 Some think that whether or not the terms 'compensation' and 'liability' are included in the agreement, the WIM or the parties to the Paris Agreement are not precluded from agreeing over time to a legal regime, which might resemble a liability scheme, or which may provide some kind of monetary payout or financial support in case of actual damage (Kreienkamp and Vanhala 2017, 206).
19 Reparations also apply to relationships among individuals or groups and do not require that these individuals or groups have had a successful relationship in the past. A repaired relationship can be a relationship that has come to good terms, even if it had never been so previously.
20 Although Butt (2009) is not as explicit in his definition of 'rectification', I believe that my account here shares his understanding of this term.
21 Throughout this book, I may use the term *prima facie* moral considerations to refer to moral considered judgments. There are different understandings of what constitutes *prima facie* moral considerations. *Prima facie* moral considerations might be regarded as providing reasons that apply provisionally when we do not know many details of the situation at hand, but 'have no residual reason-giving force' when we learn more about the situation (Hurley 1989). In this sense, *prima facie* moral considerations are 'mere appearances'. That is, in this understanding, *prima facie* moral considerations are considerations that seem to be morally relevant at first glance but, when we look into the details, are in fact not morally relevant at all. This is not how the concept of *prima facie* moral considerations is understood in this book. Instead, I follow Thomson in understanding *prima facie* moral considerations as considerations that are favorable to the attribution of certain moral duties (Thomson 1990). When I say that those who cause

harm to others should *prima facie* bear the burdens associated with repairing the harm, I mean that having caused harm to a person is a consideration relevant to obliging that person to bear at least some of the burdens associated with repairing that harm.

22 One could argue that this description is closer to narrow reflective equilibrium. The difference between narrow and wide reflective equilibrium is essentially that narrow reflective equilibrium only involves the coherence between considered judgments and moral principles, whereas wide reflective equilibrium also involves background theories. This concern might be justified if it arises from the fact that I do not include enough background theories, or even that my work is too focused on working out the coherence between considered judgments and moral principles. I would respond to such a concern with the following remarks. First, it is of course impossible to consider all relevant background theories here. Nevertheless, this work relies significantly on background theories related to the science of climate change. In my view, that sets us on the path toward wide reflective equilibrium, which is the most I can achieve in the scope of this work. Second, it is in any case debated whether full wide reflective equilibrium can be achieved. Even Rawls himself remained neutral as to whether a state of reflective equilibrium can actually be reached (Rawls 1971, 49). Thus, unless every work claiming to use reflective equilibrium is to be disregarded for not reaching a wide reflective equilibrium, the fact that my work is limited in the use of background theories should not be considered a major problem.

23 I say 'in principle' because I would like to leave the door open to the fact that, in practice, it might be difficult to disentangle the effects of climate change and the effects of other injustices. However, in principle, it is possible, and, in my view, this is what should characterize views on climate justice. Distinguishing these considerations 'in principle' allows us to answer the objection that climate scholars seek to solve all injustice by tackling climate change (Posner and Weisbach 2010). If we make this in-principle distinction, that concern can be softened.

24 The degree of controversy may vary depending on the duties assigned to those agents. Generally, backward-looking approaches might come across as more problematic (Smiley 2017).

25 Drawing from Miller, 'we can speak of the German nation, for instance, persisting through the Weimar Republic, the Nazi regime, the Federal Republic and the German Democratic Republic, and so forth' (Miller 2004, 243).

26 More recently, Francis (2020) has adopted this approach.

27 There might be a third alternative for giving an account of current duties of justice related to past actions. That third alternative would take individuals as the relevant units of concern but would not consider their duties as members of certain transgenerational groups. Instead, it would consider their duties with respect to the past in light of certain relationships with other past individuals that need not be mediated by their belonging to the same group. I will not explore this possibility any further here.

28 Some of the ideas that follow until the end of this section appear in García-Portela (2019).

29 These two strategies also allow us to circumvent or answer the Dead-Polluters Objection (García-Portela 2019). Some scholars place importance on the fact that, while many emissions have been produced by people alive today, an important amount of emissions were produced by past individuals who are no longer here (Meyer and Roser 2010; Caney 2006; Baatz 2013; Meyer 2013). One might think that assigning rectificatory duties to present-day people because of what past generations did is unfair. This has become known as the Dead-Polluters Objection. The One Entity Strategy circumvents this objection because whether some of the past members of a given nation are no longer alive is not problematic for assigning rectificatory duties to the collective entity that has survived across time. By changing the entity that is the target of rectificatory duties (from the individual to the collective level), the One Entity Strategy would annul the

concerns of the Dead-Polluters Objection. The Two Groups Strategy answers this objection by explaining why it is not unfair that people alive today bear the duties of past members of their communities.

30 Note that the strength of this concern should be softened by the fact that people alive today have also emitted quite a significant amount of emissions.

31 In this point, my approach is very similar to the one put forth by Heyward (2010), who claims that the PPP and the Ability to Pay Principle (understood in sufficientarian terms) should be combined to select duty-bearers. Thus, those who fall within the intersection between both principles are identified as the duty-bearers of climate justice.

32 This is at least the case for the international context, that is, when the duty-bearers and the victims do not belong to the same nation. In these cases, further considerations of distributive justice (such as egalitarian or prioritarian considerations above the threshold) are not strong enough to outweigh rectificatory justice duties. This assumption is justified because the duty to uphold an international sufficientarian pattern of distribution is grounded in a relationship in which human beings stand toward each other by virtue of sharing certain basic characteristics and a single world (Miller 1997). I take this to be the relevant context because most losses and damages (as described in a minimal sense in this book) constitute transboundary harm. I acknowledge that, in contexts where duty-bearers and victims share other common features (for instance, as members of the same national or cultural community, or as members of the same family, or simply belonging to the same political community or being geographically close to one another), the conclusions concerning the strength of other distributive justice considerations might be different.

33 For instance, by adopting as pluralist sufficientarian view, such as those laid out by Shields (2020).

34 However, if one opts for this alternative and places higher importance on distributive justice considerations above the threshold, this view will neglect the attribution of duties to some countries (such as China) that are historically high polluters but whose welfare levels are not as high as other historically high-polluting countries.

35 Note that my commitment to rectificatory climate justice does not involve a commitment to a sort of libertarian theory of global justice. I do not claim that we should conceive of just any type of justice in terms of rectificatory justice. I maintain only that climate justice should be understood in terms of rectificatory and should always be limited by a background sufficientarian theory of distributive justice. This means that, for those who would be left below the threshold as a result of the application of rectificatory climate justice, principles of global distributive justice should be applied to bring them to a threshold of sufficiency. This last move, however, would not necessarily qualify as a requirement of climate justice, but rather as a requirement of international and/or intergenerational justice.

References

Anderson, Benedict. 2006. *Imagined Communities: Reflections on the Origin and Spread of Nationalism*. Revised Edition. New York: Verso.

Baatz, Christian. 2013. 'Responsibility for the Past? Some Thoughts on Compensating Those Vulnerable to Climate Change in Developing Countries'. *Ethics, Policy & Environment* 16 (1): 94–110.

———. 2016. *Compensating Victims of Climate Change in Developing Countries Justification and Realization*. Dissertation, Kiel University.

Butt, Daniel. 2009. *Rectifying International Injustice: Principles of Compensation and Restitution between Nations*. Oxford; New York: Oxford University Press.

———. 2017. 'Historical Emissions. Does Ignorance Matter'? In *Climate Change and Historical Emissions*, edited by Lukas H. Meyer and Pranay Sanklecha. Cambridge: Cambridge University Press.

Calliari, Elisa. 2018. 'Loss and Damage: A Critical Discourse Analysis of Parties' Positions in Climate Change Negotiations'. *Journal of Risk Research* 21 (6): 725–47. https://doi.org/10.1080/13669877.2016.1240706.

Caney, Simon. 2006. 'Environmental Degradation, Reparations, and the Moral Significance of History'. *Journal of Social Philosophy* 37 (3): 464–82.

———. 2018. 'Distributive Justice and Climate Change'. In *The Oxford* Handbook *of Distributive Justice.*, edited by Serena Olsaretti. Oxford; New York: Oxford University Press.

Casal, Paula. 2007. 'Why Sufficiency Is Not Enough'. *Ethics* 117 (2): 296–326. https://doi.org/10.1086/510692.

Coleman, Jules. 1995. 'The Practice of Corrective Justice'. In *Philosophical Foundations of Tort Law*, edited by David Owen, 53–72. Oxford; New York: Oxford University Press.

Daniels, Norman. 1996. *Justice and Justification: Reflective Equilibrium in Theory and Practice*. Cambridge Studies in Philosophy and Public Policy. Cambridge: Cambridge University Press. https://doi.org/10.1017/CBO9780511624988.

Francis, Blake. 2020. 'In Defense of National Climate Change Responsibility: A Reply to the Fairness Objection'. *Philosophy &* Public Affairs 49 (2).

Frankfurt, Harry. 1987. 'Equality as a Moral Ideal'. *Ethics* 98 (1): 21–43.

Friman, Mathias, and Mattias Hjerpe. 2015. 'Agreement, Significance, and Understandings of Historical Responsibility in Climate Change Negotiations'. *Climate Policy* 15 (3): 302–20.

García-Portela, Laura. 2019. 'Individual Compensatory Duties for Historical Emissions and the Dead-Polluters Objection'. *Journal of Agricultural and Environmental Ethics* 32 (4): 591–609.

Gardiner, Stephen M., and David A. Weisbach. 2016. *Debating Climate Ethics. Debating Ethics*. New York: Oxford University Press.

Gleick, Peter H. 2014. 'Water, Drought, Climate Change, and Conflict in Syria'. *Weather,* Climate*, and Society* 6 (3): 331–40.

Heyd, David. 2017. 'Climate Ethics, Affirmative Action and Unjust Enrichment'. In *Climate Change and Historical Emissions*, edited by Lukas Meyer and Pranay Sanklecha, 22–45. Cambridge; New York: Cambridge University Press.

Heyward, Clare. 2010. *Environment and Cultural Identity: Towards a New Dimension of Climate Justice*. Dissertation. Oxford University.

Hurley, Susan L. 1989. *Natural Reasons:* Personality and Polity, Vol. 41. New York: Oxford University Press.

Ivison, Duncan. 2006. 'Historical Injustice'. In *The Oxford Handbook of Political Theory*, edited by John Dryzek, Bonnie Honig, and Anne Phillips. Oxford: Oxford University Press.

Kreienkamp, Julia, and Lisa Vanhala. 2017. 'Climate Change Loss and Damage'. In *Policy Brief.* Brussels: Global Governance Institute.

Mechler, Reinhard, et al., eds. 2019. *Loss and Damage from Climate Change*. Cham, Switzerland: Springer.

Meyer, Lukas. 2013. 'Why Historical Emissions Should Count'. *Chicago Journal of International Law* 13: 597.

Meyer, Lukas, and Dominic Roser. 2010. 'Climate Justice and Historical Emissions'. *Critical Review of International Social and Political Philosophy* 13 (1): 229–53.

Miller, David. 1997. *On Nationality*. Oxford; New York: Oxford University Press.

———. 2004. 'Holding Nations Responsible'. *Ethics* 114 (2): 240–68.

———. 2007. *National Responsibility and Global Justice*. Oxford; New York: Oxford University Press.

Nozick, Robert. 1974. *Anarchy, State, and Utopia*. New York: Basic Books.

Page, Edward A. 2012. 'Give It up for Climate Change: A Defence of the Beneficiary Pays Principle'. *International Theory* 4 (02): 300–30.

Page, Edward A., and Clare Heyward. 2016. 'Compensating for Climate Change Loss and Damage'. *Political Studies* 65 (2): 356–72.
Parry, M. L., O. F. Canziani, J. P. Palutikof, and P. van der Linden. 2007. *Climate Change 2007 – Impacts, Adaptation and Vulnerability: Working Group II Contribution to the Fourth Assessment Report of the IPCC*. Cambridge; New York: Cambridge University Press.
Posner, Eric A., and David A. Weisbach. 2010. *Climate Change Justice*. Princeton: Princeton University Press.
Rawls, John. 1971. *A Theory of Justice*. Original Edition. Cambridge: Belknap Press.
Shields, Liam. 2020. 'Sufficientarianism'. *Philosophy Compass* 15 (11): 1–10.
Smiley, Marion. 2017. 'Collective Responsibility'. In *The* Stanford *Encyclopedia of Philosophy*, edited by Edward N. Zalta, Summer. Stanford: Metaphysics Research Lab, Stanford University. https://plato.stanford.edu/archives/sum2017/entries/collective-responsibility/.
Thompson, Janna. 2002. *Taking Responsibility for the Past: Reparation and Historical Injustice*. Cambridge; Malden: Polity.
———. 2006. 'Collective Responsibility for Historic Injustices'. *Midwest Studies in Philosophy* 30 (1): 154–67.
———. 2017. 'Historical Responsibility and Climate Change'. In *Climate Change and Historical Emissions*, edited by Lukas H. Meyer and Pranay Sanklecha, 46–60. Cambridge; New York: Cambridge University Press.
Thomson, Judith Jarvis. 1990. *The Realm of Rights*. Cambridge: Harvard University Press.
Tommie, Shelby. 2013. 'Racial Realities and Corrective Justice'. *Critical Philosophy of Race* 1 (2): 145. https://doi.org/10.5325/critphilrace.1.2.0145.
UNFCCC. 2007. 'Report of the Conference of the Parties on Its Twenty-Eighth Session, Held in the United Arab Emirates from 30 November to 13 December 2023'. *FCCC/CP/2023/11/Add.1*
———. 2011. 'Report of the Conference of the Parties on Its Sixteenth Session, Held in Cancun from 29 November to 10 December 2010. Addendum. Part Two: Action Taken by the Conference of the Parties at Its Sixteenth Session'. *FCCC/CP/2010/7/Add.1*
———. 2014a. 'Report of the Conference of the Parties on Its Nineteenth Session, Held in Warsaw from 11 to 23 November 2013'. *FCCC/CP/2013/10/Add.3*
———. 2014b. 'Report of the Executive Committee of the Warsaw International Mechanism for Loss and Damage Associated with Climate Change Impacts'. *FCCC/SB/*2014/*4*. Lima.
———. 2015, November 30 – December 13. 'Report of the Conference of the Parties on Its Twenty-First Session, Paris'. *FCCC/CP/2015/10*.
———. 'website'. *Warsaw International Mechanism for Loss and Damage Associated with Climate Change*. https://unfccc.int/topics/adaptation-and-resilience/workstreams/loss-and-damage-ld/warsaw-international-mechanism-for-loss-and-damage-associated-with-climate-change-impacts-wim.
Zellentin, Alexa. 2015. 'Compensation for Historical Emissions and Excusable Ignorance'. *Journal of Applied Philosophy* 32 (3): 258–74.

2 A minimal capabilities-based approach

Many climate justice scholars agree that, up until now, political agreements concerning loss and damage have been made possible because of the use of ambiguous and vague language in the use of these concepts (Boyd et al. 2017; Mechler et al. 2020; Puig 2022). Recently, Friederike Otto has confirmed this intuition:

> As long as loss and damage is not clearly defined, the people causing climate change can sign agreements that will at least move the issue a little further forward. Developing countries won't see any money from it immediately, but the mere fact that loss and damage has become a separate pillar in the architecture of climate negotiations, and is thus being acknowledged, is a provisional success and a foundation upon which to build.
>
> (Otto 2018, 23)

However, advancing the implementation of loss and damage political measures requires major clarity. Notably, in an influential piece, Page and Heyward argued that 'it is clear that a major stumbling block to further progress in this arena is a series of gaps in our understanding of the meaning, application and justification of the concept of loss and damage' (Page and Heyward 2016, 3). Remarkably, since the publication of their paper, not much progress has been made.

With the agreement of creating a specific fund for loss and damage reached in 2022 by the UNFCCC parties at the last COP27 held in Egypt (UNFCCC 2022) and confirmed at the COPE28 in Dubai, the need for a clear definition of loss and damage becomes now even more pressing. This definition has two very important consequences for developing a loss and damage fund since it determines, first, to whom the funds will flow and, second, what kind of measures will be funded. The aim of this chapter is to clarify these issues before we get into questions about responsibility and implementation to contribute to the fund.

It is not that no definitions are available. For example, the UNFCCC has defined loss and damage as 'the actual and/or potential manifestation of impacts associated with climate change in developing countries that negatively affect human and natural systems' (UNFCCC 2012, 4). However, as Page and Heyward argued, this definition lacks conceptual clarity. Some of the problems are that, in the UNFCCC discourse, loss and damages are treated broadly as 'the adverse impacts of climate

DOI: 10.4324/9781003399889-2

change' (Page and Heyward 2016, 3), but it is not clear whether the impacts that count as loss and damage are the material and physical impacts of climate change (e.g., a building destroyed by a storm or a drowning island) or the impacts of climate change on people's lives (e.g., lack of adequate shelter or human displacement). Moreover, there has not been in-depth work on how the concepts of 'loss' and 'damage' differ from each other, and how the corresponding policy responses should be different. Furthermore, it is also unclear why such a definition needs to be limited to vulnerable and developing countries (Surminski and Lopez 2014).

In a nutshell, this chapter argues that loss and damage occur, *at least*, when climate change disrupts people's lives by pushing them below a sufficient standard in their opportunity to enjoy the central aspects of a dignified, flourishing life (or capabilities). I call this view the minimal capability-based account of loss and damage. Moreover, it situates loss and damage measures in relation to other climate change-related measures and proposes reparative measures to address different manifestations of loss and damage.

2.1 Life disruptions as harm and the minimal understanding of loss and damage

To solve some ambiguities, Page and Heyward define loss and damage as 'the unjustified disruptions in the lives of individuals and communities, whether permanent or otherwise, that are attributable to anthropogenic climate change and which remain after mitigation and adaptation efforts have been attempted' (Page and Heyward 2016, 3). Two things are worth highlighting here. First, for an impact to count as loss and damage and thus be addressed by L&D climate policies, there needs to exist some kind of connection between the impact and climate change (see also García-Portela 2023, 374). Second, Page and Heyward understand the normatively relevant 'currency of disruption' in loss and damage as being 'human ends', rather than the material and physical impacts of climate change. Here, they follow Amartya Sen's maxim according to which resources matter from a normative perspective for the way they allow people to do or to be (Sen 1999). Hence, from a normative perspective, loss and damage should be understood as disruptions caused by climate change on what people can do or can be because that is what matters fundamentally. In this book, I adopt this general view. However, some more conceptual work needs to be done in order to flesh out and specify how disruptions and human ends ought to be understood in the context of loss and damage as well as the various forms they might take. This section starts my analysis by delving into the concept of life disruption. In Chapter 5, I will address how we can establish a connection between life disruptions and climate change so that we can speak about them as loss and damage from climate change. For reasons that will become obvious in a moment, I will refer to this as the 'identification of harm' problem.

Life disruptions can occur in different ways. One's life can be disrupted by an overall positive change. For instance, one's life can be disrupted by becoming a parent. Despite being often reported as an overall positive experience, becoming a parent is an abrupt change that has many implications for how one lives one's life.

However, most often the term 'disruption' refers to a harmful impact on someone's life. That is, an individual's life is disrupted by a certain event if the individual is harmed by such an event, in the sense of being made worse off than they would have otherwise been in the absence of such an event. Without doubt, this is the way in which 'disruptions' are used in the context of loss and damage from climate change. Thus, the concept of loss and damage relies on the notion of life disruption and this one, in turn, on the notion of (climate) harm.

However, this notion of harm is controversial in intergenerational contexts such as climate change because of the well-known non-identity problem (Parfit 1984). Past activities responsible for climate change (the use of trains, cars, planes or the reliance of our energy systems on fossil fuels, etc.) affect the standard of living of presently living people. Moreover, because of their impact on social interactions, they also affected who will come into existence. However, it is difficult to claim that the activities that bring people into existence also harm them in the sense of making them worse off than they would have otherwise been (counterfactual notion of harm). Presently living people affected by past climate change–inducing activities are not worse off than they would have otherwise been. Instead, in the absence of climate change–inducing activities, they would have never existed. If this is true, it is likely that members of the current generation have not been made worse off, and thus harmed, by climate change (Page 2008; Caney 2006).

Does this mean that people cannot suffer harm in the form of life disruptions due to climate change? I do not think so. Notice that the non-identity problem rests on a counterfactual notion of harm. But a different notion of harm might be able to make sense of the idea that our actions can harm future people or that currently living people have been harmed by the actions of previous generations. Some authors have argued that a sufficientarian threshold notion of harm could indeed circumvent the non-identity problem. According to a sufficientarian threshold notion of harm, a person is harmed when they are pushed below a level sufficient to live a decent and dignified flourishing life, thereby coming to live a life with intrinsically bad properties (Meyer and Roser 2009; Meyer 2015).

The threshold notion of harm circumvents the non-identity problem because it is non-comparative; that is, it does not require comparing the state of a person to a counterfactual one. When we want to find out whether a person has been harmed by an event (in this case, climate change), we do not need to research what her state would have been in the absence of that event. Instead, it is enough to find out what her current circumstances are because of that event. For a person to suffer harm, it is enough that they are made to fall below a sufficiency threshold. Arguably, many people who suffer the negative consequences of climate change have been harmed in this threshold sense, at least intuitively. Those who live under the permanent threat of being displaced by sea-level rising or those who lack enough food due to the exacerbation of drought events seem to fall under a threshold of a flourishing, dignified life, whose intrinsically bad properties we will explore in the next section. The minimal capabilities-based approach to loss and damage is based on the threshold notion of harm.

One might not be convinced about the relevance of the non-identity problem and think that this is only a philosophical puzzle with no real-world relevance

(Butt 2009). However, even in this case, understanding loss and damage along the sufficientarian threshold notion of harm can play an additional role. Even if one disregards the non-identity problem and believes that harm can occur in intergenerational contexts in a counterfactual sense, one could still doubt whether all the negative impacts of climate change should be compensated for, or whether they should also have the same normative status. Think about the multimillionaire who loses his twentieth house in a flood event close to the coastline. In this case, this multimillionaire might suffer counterfactual harm, bracketing the non-identity problem. But, arguably, repairing loss and damage might not be a priority because it does not undermine the multimillionaire's life in a significant enough way, and it can be very costly for society to bear such a burden. One might argue that this money could be used instead for other climate change-related purposes, such as further mitigation or adaptation projects, or even for other societal purposes. Taking this into consideration, one might question whether such negative effects should be covered by, or at least be the priority of, loss and damage policies.

I do not intend to commit here with such a view. However, the sufficientarian threshold notion of harm can identify at least a minimal way in which loss and damage can occur and the kind of climate harm that should be the focus of and have priority in L&D policy. Such a minimal definition could provide the grounds for a wide agreement about what counts as loss and damage. In my view, stakeholders could agree that loss and damage occur, at least, when people are pushed below a level sufficient to live a decent and dignified flourishing life and that L&D policies should be focused on and give priority to this kind of climate harm.

Moreover, the notion of threshold harm involved in this minimal account justifies the focus on developing and vulnerable countries in the loss and damage discourse, even if we exclude this reference from the general definition. We can expect that it is in such countries that the negative effects of climate change impact people's ability to enjoy the minimum to live a sufficiently decent life. In other words, my account clarifies that, even though it would be unjustified to limit loss and damage to impacts occurring in vulnerable and developing countries, loss and damage thusly defined are more prone to occur in those countries and therefore a particular policy focus on them is justified.

2.2 A minimal capabilities-based account of loss and damage

According to a sufficientarian threshold notion of harm, a person is harmed when they are pushed below a level sufficient to live a decent and dignified flourishing life, thereby coming to live a life with intrinsically bad properties. When a person is harmed in this way due to climate change, they experience loss and damage. However, to flesh out that definition, we need to specify what these intrinsically bad properties are. Here, I offer an interpretation along the lines of the capabilities approach to describing loss and damage.

The capabilities approach states that there are certain objective and universal core elements of people's flourishing life (functionings) whose opportunities for realization (capabilities) should be guaranteed and not thwarted as a matter of social justice (Sen 2009; Nussbaum 2007, 2011). Note that the relevant notion is the one of 'opportunities', capabilities, to achieve certain functionings and not the achievement of the functionings themselves. Someone can enjoy a perfectly flourishing life without achieving these functionings provided that they have the real opportunity to do so. The classical example (Sen 2009, 237) is that of a person who voluntarily fasts (even though they might not reach a certain level of functionings, the fact that they can do so leaves their capabilities intact). According to my proposal, people are harmed in this threshold sense when they are pushed below a sufficientarian level in at least one of their central capabilities, and we can describe the harmful impacts involved in loss and damage in these terms.

Famously, Nussbaum has provided a list of central capabilities (Nussbaum 2007, 2011).[1]

Box 2.1 Nussbaum's capabilities list (2007, 2011)

1. Life: . . . not dying prematurely, or before one's life is so reduced as to be not worth living.
2. Bodily health: Being able to have good health . . .; to be adequately nourished; to have adequate shelter.
3. Bodily integrity: Being able to move freely from place to place . . .
4. Senses, imagination and thought: . . . Being able to use imagination and thought in connection with experiencing and producing works and events of one's choice . . .
5. Emotions. Being able to have attachments to things and people; . . . to love, to grieve, to experience longing, gratitude, and justified anger. Not having one's emotional development blighted by fear and anxiety. . . .
6. Practical reason. Being able to form a conception of the good and to engage in critical reflection about the planning of one's life. . . .
7. Affiliation. Being able . . . to engage in various forms of social interaction; . . .
   Having the social bases of self-respect . . .; being able to be treated as a dignified being whose worth is equal to that of others. . . .
8. Play: Being able to laugh, to play, to enjoy recreational activities.
9. Other species: Being able to live . . . in relation to animals, plants, and the world of nature.
10. Control over one's environment.
    A. Political. Being able to participate effectively in political choices that govern one's life; . . .
    B. Material. Being able to hold property [and] having the right to seek employment on an equal basis with others.

If a person's capability is pushed below a sufficiency level, this constitutes an intrinsically bad property for the life of this person because it prevents the person from enjoying a decent and flourishing life, or, as Nussbaum would put it, a truly human life or a life of dignity (Nussbaum 2007, 71). Having one or various capabilities below this sufficientarian threshold is intrinsically bad for people and, in this sense, constitutes threshold harm. Consequently, loss and damage occur when, because of the negative effects of climate change, people experience life disruptions that push them below a sufficient level of certain capabilities.

Capabilities theorists are very aware that their philosophical approach may not permeate the political discourse. Instead, the language of rights may be more suitable for this task. Nussbaum seems to be very aware of this fact when she writes:

> When governments and international agencies talk about people's basic political and economic entitlements, they regularly use the language of rights. When constitutions are written in the modern era, and their framers wish to identify a group of particularly urgent interests that deserve special protection, once again it is the language of rights that is regularly preferred.
> (Nussbaum 1997, 273)

The capabilities approach can be thusly also understood as a type of human rights approach, as Nussbaum herself believes (Nussbaum 1997). Each of those capabilities represented in the capabilities list are minimal entitlements and, accordingly, correspond to a human right. Human rights grounds people's entitlements to having a certain level of capabilities protected and guaranteed. Following the sufficientarian approach that I have embraced here, this level would be, at least minimally, a sufficiency one.

This connection between the capabilities and human rights shares some features with standard understanding of human rights in the context of climate change. Prominently, Simon Caney has argued that human rights have four core features. First, human rights are grounded *in each person's humanity*. Human beings have certain rights because of their shared characteristics as beings of certain type. One could refer here to different set of morally relevant characteristics that human possess and make them worth of having those rights. In my account, these features could be the functionings related to each of the capabilities. Second, human rights represent entitlements to certain minimal standards of treatment, and they generate obligations on all persons to respect these basic minimum standards (Caney 2009, 165). In my account, these basic standards would be sufficientarian ones. Third, human rights take priority over other values. Fourth, moral thresholds: they represent levels below which individuals should not be permitted to fall. That is, human rights would ground a distributive justice duty to guarantee that people have enough for living a decent and dignified life and they would ground a duty not to make people fall below this threshold. In a nutshell, following the standard account provided by Caney, 'human rights specify minimum moral thresholds to which all individuals are entitled, *simply in virtue of their humanity*, and which override all other moral values'. (Caney 2009, 165). In my account, those are sufficientarian

Table 2.1 Typology of climate change measures

Temporal dimension	*Ex ante*		*Ex post*
Climate policies	Mitigation	Adaptation	L&D
Aim	Avoiding climate change pushing people's capabilities below a sufficiency threshold		Repairing the negative effects of climate change on people's sufficiency threshold of capabilities/human rights
Method	Reducing CO_2 emissions or improving carbon sinks	Reducing vulnerabilities or enhancing coping mechanisms	Material reparations and symbolic reparations

thresholds of capabilities. Then, according to what I have put forward here, loss and damage occur when, because of the negative effects of climate change, people experience life disruptions that push them below a sufficient level of certain capabilities, thereby also infringing their human rights.

With this view in mind, we can differentiate between mitigation and adaptation as ex ante or prospective policies, on the one hand, and loss and damage as ex post or retrospective policies, on the other hand. Mitigation policies aim to avoid climate change pushing people's capabilities below a sufficiency threshold by reducing the total amount of emissions in the atmosphere and enhancing carbon sinks. This amounts to say that mitigation policies aim to protect people's human rights. Adaptation policies also aim to avoid people's capabilities falling below a sufficiency threshold due to the negative effects of climate change, but they do so by adjusting populations to the observable negative effects of climate change or to those threats that are foreseeable in the near future. Or, in other words, adaptation policies aim to react to and protect from the foreseeable infringement of people's human rights. Finally, L&D policies aim to repair the negative effects of climate change on people's capabilities that cause them to fall below a sufficiency threshold with material and symbolic reparations. In other worlds, L&D measures aim are repairing infringements on people's human rights. In this way, L&D policies have an eminent reparatory aim.

These temporal distinctions take the effects of climate change on people's sufficiency level of each capability as the reference point. Table 2.1 represents these distinctions.

2.3 Answering some challenges to an ex post categorization of L&D

My ex post categorization of L&D could be challenged at least on two fronts. These challenges arise from a potential conflict between my proposed definition and how L&D measures have been framed broadly in the political and academic discourse.

Answering these two challenges here allows me to justify the retrospective dimension of loss and damage and develop my capabilities-based account further.

The first challenge concerns the inclusion of risk management measures within L&D measures. Various scholars have argued that L&D include at least some risk management measures in response to climate change. They differentiate between acceptable, tolerable, and intolerable risks (Dow et al. 2013) and argue that L&D measures deal with intolerable risks (Wallimann-Helmer 2015; Reinhard Mechler and Schinko 2016; Wallimann-Helmer et al. 2018; Schinko, Mechler, and Hochrainer-Stigler 2018). However, risk management of intolerable risks occurs ex ante, that is, before the impacts of climate change materialize. But if this kind of risk management is an ex ante measure and is considered part of L&D measures, the category of L&D cannot be considered fully ex post.[2]

In response to this challenge, I argue that my approach is compatible with the inclusion of the management of intolerable risks within L&D without undermining the categorization of L&D as ex post measures. In a nutshell, this is because intolerable risks are precisely defined as risks that already affect people's ability to reach a sufficient level of capabilities.

A clear definition of 'intolerable risks' is lacking in the literature. One of the best attempts has been made by Wallimann Helmer, who claims that 'intolerable risks arise when a human system is not able to adapt to anticipated negative climate impacts' (Wallimann-Helmer 2015, 472). Unfortunately, this definition is circular. Note that L&D measures are measures to be undertaken when adaptation is no longer possible. If intolerable risks are defined by reference to adaptation, that definition implies that what distinguishes L&D measures from adaptation is that L&D measures concern intolerable risks and that intolerable risks are those risks that adaptation does not cover. The question that still remains is how to describe the kind of risks to which one is no longer able to adapt, that is, intolerable risks.

Wallimann-Helmer provides two claims that can help in providing an answer to that question. First, he claims that 'the limits of adaptation are reached when risks become intolerable' (Wallimann-Helmer 2015, 472). Second, he claims that 'the limits of adaptation are reached when a human system is no longer able to secure valued objectives' (ibid). These two claims seem to suggest that an intolerable risk is reached when a human system can no longer secure valuable objectives. Naturally, the next question is what those 'valuable objectives' are. Although Wallimann-Helmer does not clarify this last point, his reliance on Dow et al. (2013) might provide adequate guidance. These scholars exemplify the notion of limits to adaptation and intolerable risks thus:

> A farmer seeking to cultivate a specific crop under increasingly stressed water resource will invest in . . . increasing adaptive effort as access to water resources becomes more constrained . . .. *But, at some point, no new adaptation options are available to respond to growing risks, or the level of adaptive effort required to maintain valued objectives becomes infeasible.* At this point, the farmer may, for example, choose to abandon farming altogether.
>
> (Dow et al. 2013, 306; italics mine)

Following this example, it looks as if the farmer's valuable objectives are attached to their life as a farmer. This could include their work as a farmer and the forms of affiliation and social interaction associated with their farming activities. Or perhaps their close relation to nature. However, due to water stress, continuing their life as a farmer, although a valuable objective, is too risky. For instance, one could imagine that their life as a farmer is at risk because, if they continue, they might lose too much money. Or they might not have enough food to eat next month, given that he will not be able to earn enough food from their farming. Arguably, this is too much of a risk to run. Or, in other words, this is an intolerable risk because they can no longer secure his valuable objectives. Because of the risks they face, they need to stop farming, thereby abandoning one of those valuable objectives. Probably, also, they need to search for a different job and perhaps even in a different location.

Notice that this same idea can be described in terms of the capabilities approach and also in terms of the ex post understanding of L&D. Arguably, these valuable objectives are the central elements of a person's flourishing life or functionings. In fact, the valuable objectives mentioned before map quite nicely with the functionings corresponding to some capabilities mentioned earlier, for instance, the capabilities of enjoying our relationship with other species and particularly with the world of nature, affiliation, practical reason, or control over one material and political environment (see Figure 6.1). But because of the risk of, for instance, going bankrupt or not having to eat next month, the farmer has to abandon their activity, which means that they no longer have the real opportunity to develop those core elements of the flourishing life to a sufficient level. And the real opportunities to develop those core elements of a flourishing life are precisely the capabilities. That is, they cannot enjoy their capabilities.

Of course, the farmer might be able to change timely to a new profession. This new profession might be able to satisfy some of the affected capabilities and to do so at a sufficient level. In this case and for those capabilities, the farmer would be able to *adapt* to the negative consequences of climate change without experiencing loss and damage. However, it is also plausible that this would not be the case for all the capabilities affected by climate change and, thus, at least in some aspects of their life, the farmer might suffer loss and damage. Hence, the aforementioned example may involve adaptation and loss and damage dimensions.

In other words, the farmer experiences loss and damage *in some aspects of their life* precisely when they can no longer enjoy *certain* capabilities (valuable objectives) at a sufficient level because of the risk they face. Hence, the inclusion of risk management of intolerable risks within L&D measures does not challenge my ex post understanding of L&D. Instead, L&D measures should be understood as ex post measures that include the management of those kinds of risks that already affect people's enjoyment of capabilities at a sufficiency level.

The second challenge concerns capacity-building for implementing L&D measures. Arguably, the capacity for responding to L&D needs to be in place before loss and damage occur. This might suggest that at least some L&D measures must be established ex ante (long) before climate impacts actually materialize (Wallimann-Helmer 2015, 471). For instance, L&D measures might include insurance schemes that must be established before climate change impacts occur (Mechler and Schinko 2016; Linnerooth-Bayer et al. 2019).

However, note that my approach can make sense of the inclusion of insurance schemes in the sphere of L&D without undermining my ex post categorization. Here, we could differentiate between the establishment of certain mechanisms and the gathering of certain resources (i.e., such as insurance policies) and L&D measures themselves. Whether capacities are built and resources (e.g., money) are gathered beforehand, the relevant L&D measures concern such matters as providing money job training opportunities, and relocation expenses. But notice that these measures are not undertaken before loss and damage occur. Although the capacities to carry out those particular measures might be established beforehand, the actual restoration of capabilities occurs ex post. In my view, it is not the insurance mechanism itself that is the L&D measure but the particular actions that are carried out using the funds of the insurance policy to address loss and damage. This is better described as an ex post L&D measure that involves an insurance mechanism.

2.4 Conceptual clarifications and types of reparation for loss and damage

In this section, I propose an interpretation of the two concepts 'loss' and 'damage' following the minimal capabilities-based approach. Moreover, I do the same with the concepts of economic and non-economic losses and damage, and the kind of reparations corresponding to these categories.

2.4.1 The notions of 'loss' and 'damage' in loss and damage

It is not uncommon to find 'loss' and 'damage' as synonyms in the literature on loss and damage (Preston 2017). The UNFCCC refers to 'losses' as 'negative impacts in relation to which reparation or restoration is impossible' and 'damages' as 'negative impacts in relation to which reparation or restoration is possible' (UNFCCC 2012, 3). Some have interpreted the distinction as referring to 'loss' as a permanent disruption to human lives and 'damage' as disruption that is reversible (Kreft, Harmeling, and Warner 2012; Huq, Roberts, and Fenton 2013).

The capabilities-based framework draws and elaborates on this distinction. To begin, recall from the first section that the relevant impacts that concern loss and damage are those that compromise the sufficiency level of people's capabilities. For that reason, my analysis here uses the definition of loss and damage as concerning the effects of climate change on people's capabilities and not the effects of climate change on objects. The individual definitions of 'loss' and 'damage' follow this idea.

Following the capabilities-based framework, losses are permanent disruptions to the enjoyment of a central capability at a sufficient level, whereas damages are temporary impairments on the enjoyment of a central capability at a sufficient level. For instance, imagine that a building gets severely damaged after a storm event. The affected capability here is that of bodily health, which includes access to adequate shelter. If the inhabitants of that building can be relocated somewhere else without other capabilities, such as affiliation or practical reason, being affected, then we can speak of temporary damage in the sufficient enjoyment of their capability of bodily health. However, if they cannot be relocated without other capabilities being affected, then this temporary damage of their bodily health might be accompanied by permanent loss affecting their enjoyment of a sufficient level of a certain capability. This could happen, for instance, if these people were relocated to an area where they could not develop meaningful social interactions due to important cultural differences. In that case, those people would experience permanent loss to the sufficient level of their capability of affiliation, or their capability of control over their political environment.[3]

These definitions are in line with the UNFCCC discourse, with the caveat that the UNFCCC merely refers to loss and damage as 'impacts', whereas here I propose to read these impacts as referring to the enjoyment of capabilities at a sufficient level.

2.4.2 Reparations for economic damage, non-economic losses, and non-economic damage

This section also deals with the notions involved in loss and damage discourses, but this time the notions under scrutiny are economic and non-economic losses and damage. Although the UNFCCC distinguishes only between economic and non-economic losses, here, I argue that the relevant notions in a loss and damage capability-based framework should be the notions of economic damage, non-economic losses, and non-economic damage. Admittedly, this is a somewhat unorthodox view, but one that better serves the purposes of explaining loss and damage as harmful impacts to people's sufficient levels of capabilities. Moreover, in this section I propose a categorization not only of the harmful impacts of climate change but also of the various types of L&D reparations that are owed to those who suffer loss and damage.

The UNFCCC (2013) defines economic and non-economic losses as follows:

- Economic loss concerns 'the loss of resources, goods and services that are commonly traded in markets' (UNFCCC 2013, para. 3).
- Non-economic losses concerns 'items that are not commonly traded in markets' (UNFCCC 2013, para. 4).

In the previous section, I proposed using the terms 'loss' and 'damage' to refer to capabilities instead of objects. However, for the sake of the following argument, let us leave that framework aside for a moment and refer to loss and damage as applied to objects. These definitions seem to clash with the UNFCCC definition of loss defined as the 'negative impacts in relation to which reparation or restoration is impossible'. Notice that if losses are the impacts for which reparations or restoration are not possible, it is difficult to see how they can be economic. If an economic loss can be tradable in the market, one could restore this loss by simply quantifying its economic value and transferring that quantity of money to the person who suffered the loss. Consequently, the loss is not impossible to restore or repair, and even less can it be considered permanent. This clashes with the definition of loss provided by the UNFCCC. For reasons of coherence, because I have previously embraced and defended a particular interpretation of the use of the UNFCCC definitions of 'loss' and 'damage', I propose to avoid here the use of 'economic loss'.

Interestingly, the UNFCCC speaks of 'economic loss' but not of 'economic damage'. Here, I propose to use the notion of economic damage to express something similar to what might be expressed by the notion of 'economic loss' in the UNFCCC document. Economic damage concerns the temporary unavailability of items that can be restored or repaired, at least to a certain level, by economic means. However, since I have used the terms loss and damage in describing the capabilities approach, this definition needs to be formulated accordingly. We can define economic damage as the temporary impairment of the enjoyment of capabilities at a level that can be repaired through economic means. In my view, this notion of 'economic damage' coheres better both with the definition of loss and damage provided by the UNFCCC and with the capabilities approach presented here.

Economic damage in this sense is the object of material reparations. Material reparations concern the rectification of economic damage. This kind of reparation involves the rectification of the negative effects of climate change on people's enjoyment of capabilities at a sufficient level by economic and material means. For instance, material reparations for economic damage involve the rectification of the negative effects of climate change on housing infrastructure. Recall the example mentioned earlier. Extreme rainfall might cause severe damage to people's houses, thereby depriving them of adequate shelter. Such an event would cause economic damage by impairing people's capabilities temporarily in ways that can be repaired through economic means. For instance, such an event would impair their capability of bodily health, whose definition includes access to adequate shelter. Of course, not all loss and damage involved in such an event are economic damage in this sense. But at least some of them are, and those are the ones we refer to when we speak of material reparations for economic damage.

In a nutshell, I propose to substitute the notion of 'economic loss', described as 'the loss of resources, goods, and services that are commonly traded in markets', with the notion of economic damage, described as 'the temporary impairment in the enjoyment of capabilities at a sufficient level that can be repaired through economic means'. Moreover, I propose to use the term 'material reparations' to refer to reparations that concern economic damage. In my view, these definitions are more coherent both with the notions of loss and damage involved in my capability-based account and with the definitions provided by the UNFCCC.

Let us now turn to the notion of non-economic losses. As we saw, the UNFCCC describes non-economic losses as that concerning 'items that are not commonly traded in markets'. In the terms I propose here, we could understand non-economic losses as concerning the permanent impairment of people's enjoyment of capabilities at a sufficient level that cannot be fully repaired through economic means. In my view, this description of non-economic losses is coherent with the UNFCCC description of non-economic losses because the core idea is that these losses concern goods that cannot be reduced to their economic value. The main difference is that, instead of using the term 'loss' as applied to goods or objects, I use it here to refer to people's capabilities.

The 2013 UNFCCC technical paper 'Non-economic losses in the context of the work program on loss and damage' gives an overview of the types of non-economic loss and damage linked to climate change and the impacts they have on human lives. Notably, many of these correspond to items on the capabilities list, as shown in Table 2.2.

Unfortunately, I cannot here develop a comprehensive analysis of how these elements relate to each other, nor can I provide an exhaustive list of non-economic losses associated with capabilities. However, there are two paradigmatic cases of how non-economic loss relates to losses associated with people's central capabilities. These are the case of the Inuit Circumpolar Conference Petition and the claims of Small Island States (SIS) (Heyward 2010, 2012).[4]

The Inuit people, who live in different states around the Arctic Circle, have suffered loss and damage due to global warming. Warmer temperatures and shifts in seasonal patterns are causing ice to melt to the point of undermining the central capabilities of the Inuit people. The Inuit Circumpolar Conference petition claimed

Table 2.2 Relation between non-economic losses and damage and capabilities

Non-economic losses (UNFCCC 2013)	*Nussbaum (2007, 2011)*
Life: loss of life.	Life
Health: loss of physical and psychological health associated with respiratory diseases, cholera, sunstrokes, etc.	Bodily health
Displacement and human mobility: associated with loss of security and agency.	Bodily integrity Control over one's environment (material and political) Practical reason
Territory: loss of sovereignty and sense of place.	Affiliation Emotions Senses, imagination and thought Practical reason Control over one's environment
Cultural heritage: associated with loss of social cohesion and identity.	Affiliation Emotions Senses, imagination and thought Practical reason
Indigenous knowledge and other social capital: associated with loss of social cohesion and control over the environment.	Affiliation Emotions Senses, imagination and thought Practical reason
Biodiversity and ecosystem services: loss of diversity of living organisms and supporting, provisioning, regulating, and cultural services provided by ecosystems.	Other species Bodily health

that the effects of climate change have undercut the Inuit's ability to enjoy their ways of life and have had an important impact on their health, safety, subsistence harvest, travel, and cultural and social affiliation (ICC 2005, 67). These effects are both causing what I have called here economic damage, for instance, concerning their access to food and shelter, and non-economic losses because these effects are impacting the Inuit's survival as a distinct and unique society.

Changes brought about by rising temperatures have affected their cultural and social activities linked to food preservation, igloo building, and traditional hunting activities, which are associated with their identity as a social and cultural group. Unlike economic damage, the loss of the activities linked to their identity as a group causes permanent and irreversible loss in their enjoyment of capabilities at sufficient levels, such as their capability of affiliation and practical reason (see Figure 6.1). The loss of these capabilities constitutes a non-economic loss because these capabilities cannot be restored by material or economic means.

The case of communities living in SIS is similar. Rising sea levels due to climate change have caused floods that threaten settlements and infrastructure and severely reduce clean water availability. Many of these states, such as Kiribati and Tuvalu, are low-lying and face the risk of going completely underwater. Worsening life conditions and the threat of disappearing under water will probably cause the loss of their territory and cause them to migrate. The loss of territory has economic dimensions that will constitute temporary damage to their central capabilities. However, these can be restored with appropriate material reparations. But the loss of territory has implications for the existence of a self-governing community and the political control they can have over their territory (Bell 2004). Therefore, migrating to a different territory would come with non-economic loss of capabilities such as control over their political environment and affiliation.

The Inuit peoples and SIS cases show the kind of non-economic losses that people suffer due to climate change. These are losses because there is at least one sense in which these capabilities will be permanently impaired. They are also non-economic because they cannot be repaired through economic means. Instead, the reparations that are appropriate in these cases are victim-centered symbolic reparations. According to the account I offer here, victim-centered symbolic reparations aim to preserve the history and culture of victims and affirm the value of what has been lost. These measures are important for the sense of identity of the remaining community members who have experienced these losses.

Victim-centered symbolic reparations include measures of remembrance and commemoration initiatives that 'enable victims of climate change injustice to record their story, to recount what was lost and its effects upon them' (Heyward 2012, 163). Moreover, 'there should also be provisions made for the preservation of aspects of the group's cultural heritage: arts, technology, crafts, etc.' (Heyward 2010, 269). Similarly, Serdeczny et al. (2018) have suggested that these measures should be drawn from historical analogues of loss and practices of memorization. As Barnett et al. (2016) have argued, the ultimate goal of those practices is to manage grief and sustain the association with what would otherwise be forgotten. In this sense, museums and memorials can be thought to be among the most prominent victim-centered forms of symbolic reparations. Whereas these practices cannot make fully up for the loss of the relevant capabilities (such as affiliation and practical reason) at the sufficiency level, they are important for the sense of identity of the remaining community members who have experienced these losses and thus help in mitigating their suffering.

Remembrance and commemoration aim at preserving the cultural values that have been lost due to forced migration and the negative effects of climate change. However, forced migration causes also the loss of political self-determination, thereby negatively affecting the capability of control over one's political environment. Some other measures might be helpful not only to lessen the loss of political self-determination but also to affirm its value. For example, a free movement passport has been proposed to help the territorially disposed rebuild their political and cultural identities in a new territory, at least to the extent that this is possible (Heyward and Ödalen 2016). At the same time, giving them the choice of where to relocate and where to rebuild their cultural and political identity with a territory affirms the value of their (now lost) political self-determination and thus can be seen, too, as part of victim-centered symbolic reparations.

Victim-centered symbolic reparations should be distinguished from agent-centered symbolic reparations, which might be included to some extent at least under certain normative accounts for loss and damage (García-Portela 2020). Agent-centered symbolic reparations address non-economic damage concerning the effects of climate change on relationships of respect among differently responsible parties. As with 'economic loss', the UNFCCC does not speak of 'non-economic damage'. Yet, I believe that there might be some room for this concept.

There is a sense in which we might speak of non-economic damage, which concerns the relationships of respect between those most responsible for climate harm and those who suffer climate change-related harm. There are various normative accounts that would ground the existence of this kind of non-economic damages and thus the duty to provide agent-centered symbolic reparations by certain agents. For some of them, non-economic damage of this sort occurs when there is some kind of wrongdoing perpetrated by polluters toward climate change victims, particularly if polluters already knew about the negative effects of climate change (Heyward 2010, 2012; Page and Heyward 2016). Somewhere else, I have argued, based on the notion of moral responsibility as accountability, that the mere existence of human rights infringements of human rights is a reason for this relationship to be impaired (García-Portela 2020). I will come back to this issue in Chapter 4. For now, I am only interested in highlighting the existence of non-economic damage in this sense, regardless of its grounds. Unlike economic damage and non-economic loss, non-economic damage does not directly impair the sufficient level of any capability but occurs at a meta-level, whenever any of the capabilities are pushed below a sufficiency level because of climate change.

This impairment of the relationship of respect might be temporary, depending on the measures undertaken to repair it. In principle, the relationship between victims and responsible parties can be repaired through agent-centered symbolic reparations. Agent-centered measures may include public apologies, acknowledgments, statements of agent regret, and commitments to nonrepetition. In the context of the UNFCCC, these measures are most appropriate to be undertaken by states whose high development levels have been attached to intensive emission-generating activities and would be directed to those members of the global community who suffer the negative effects of climate change. Agent-centered measures concern thusly the self-understanding of the state undertaking these acts because the acts involve an acknowledgment of the consequences of their actions as part of their history. The purpose of these measures is to acknowledge the role of the responsible parties and eventually to repair the relations of respect among differentially responsible and affected parties. Consequently, agent-centered measures of symbolic reparations have two dimensions: one that looks backward to acknowledge certain harms as committed by the agents and one that looks forward to developing their contemporary self-understanding through the repudiation of these harms and the commitment to their nonrepetition.

Notice that agent-centered measures will not achieve their goal if the responsible parties themselves do not perform them. That is, agent-centered symbolic reparations presuppose that responsibility for having caused climate change plays at least some role in the distribution of (at least some) L&D duties. This is different from other reparations for loss and damage, which could be distributed according to other principles rather than responsibility-based principles. Here, I leave open which should be the normative guiding principles for most L&D reparations, to which I will come back in Chapters 3 and 4. If someone disagrees that responsibility for causing climate change should play a role, then one would also need to exclude agent-centered symbolic reparations in the bundle of L&D measures, as well as the idea of rectificatory climate justice. However, my aim here is merely to point out which loss and damage categories are plausible and which reparatory measures would correspond with them.

Finally, notice that economic and non-economic loss and damage are usually intertwined. Loss and damage from climate change usually have both economic and non-economic consequences. People who suffer the negative consequences of extreme precipitations and floods by losing their houses have been pushed below a threshold of their bodily health capability because they are deprived of shelter and access to clean running water. They might also suffer non-economic losses. If the consequences of the climatic event are too severe, they might be forced to move out of their villages, thereby losing control over their political environment. They might also lose their forms of affiliation. Moreover, such an event also impairs their relations of respect with those who contribute most significantly to climate change. By being pushed below a relevant threshold of capabilities, people are disrespected in ways that severely damage their relationship with those who are responsible for them. All this loss and damage is intertwined, and it often appears in single cases of loss and damage.

Table 2.3 offers an overview of the different types of loss and damage and the related reparative measures.

Table 2.3 Types of loss and damage and corresponding reparations

	Ex post measures				
Policy	*L&D*				
Aim	Repairing the negative effects of climate change on people's sufficiency threshold of capabilities				
Method	Reparations	Material reparations		Economic damage	Repairing the material effects of climate change related to people's enjoyment of capabilities at a sufficient level
		Symbolic reparations	Victim-centered	Non-economic losses	Remembrance Commemoration
			Agent-centered	Non-economic damage (affecting relations of respect)	Public apologies Statements of (agent-) regret Commitment to non-repetition

2.5 Conclusion

After the agreement reached at the COP27 in Egypt in 2022 for creating a specific fund for loss and damage, the need for a clear definition of this concept becomes a major issue for climate policymaking. Such a definition is of outmost importance because it will determine to whom the funds will flow and what measures will be funded.

This chapter has provided a definition based on the capabilities approach, thereby bridging the gap between philosophical work and the reality of policymaking. According to this definition, loss and damage refer to a particular kind of life disruptions caused by climate change, namely, those that cause negative effects of people's lives and thus harm. Loss and damage occur when climate change impacts negatively on people's enjoyment of a sufficient level of capabilities, thereby infringing their human rights. This definition not only enables us to circumvent the non-identify problem but also enables to set normative priorities when it comes to addressing the negative effects of climate change.

Following this definition, I have classified mitigation and adaptation and ex ante measures and L&D as ex post measures. Moreover, I have explained how this approach can circumvent two challenges. First, I have argued that my approach is compatible with the inclusion of the management of intolerable risks (sometimes considered as ex ante measures) within L&D without undermining the categorization of L&D as ex post measures. I have done so by challenging the idea that intolerable risks are ex ante, in the sense invoked in this book. Instead, intolerable risks are precisely defined as risks that already affect people's ability to reach a sufficient level of capabilities and thus occur already ex post, when a sufficient level of capabilities have already been compromised. Second, I have argued that even if capacity-building occurs ex ante, L&D measures refer to the provision of reparations after climatic events affect people's sufficiency level of capabilities. Hence, L&D measures remain ex post, even if they rely on previously managed capacity-building.

Finally, this chapter has offered a categorization of loss and damage following the capabilities approach. I have described losses as permanent disruptions to the enjoyment of a central capability at a sufficient level and damages as temporary impairments on the enjoyment of a central capability at a sufficient level. Then, I have differentiated between economic damage, non-economic damage, and non-economic losses. I have argued that economic damages are temporary impairments on the enjoyment of a central capability at a sufficient level that can be repaired through economic means (material reparations). Non-economic losses refer to the permanent impairment of on the enjoyment of a central capability at a sufficient level that might be repaired through victim-centered symbolic reparations. Finally, non-economic damage refers to the temporary impairment of relations of respect among different groups, which might be repaired through agent-centered symbolic reparations.

Notes

1 This specification of a central list of capabilities has led to accusations of 'westerncentrism' (Jaggar 2006). The accusation is that Nussbaum is just representing a conception of human flourishing from a Western perspective. Although it is true that capabilities are taken to be objective and universal, there is room for incorporating cultural differences in the way capabilities are realized.

2 One could also argue for the exclusion of risk management measures from L&D measures, but this would be at odds with many understandings of L&D and thus require more argumentation. Instead, I argue that risk management of intolerable risk can be explained within my capabilities-based approach.

3 Colloquially, we might also use the term 'loss' to refer to temporary disruptions and 'damage' to refer to some permanent disruptions (Page and Heyward 2016, 8). However, in my view, since the distinction between permanent and temporary disruptions is, normatively, the most important one, it is useful to align this distinction with the one between loss and damage, thereby respecting the UNFCCC terms.

4 More examples can be found in Warner and Geest (2013), Morrissey and Oliver-Smith (2013), Frankhauser, Dietz, and Gradwell (2014), and Kreienkamp and Vanhala (2017).

References

Barnett, Jon, Petra Tschakert, Lesley Head, and W. Neil Adger. 2016. 'A Science of Loss'. *Nature Climate Change* 6 (11): 976–78.

Bell, Derek. 2004. 'Environmental Refugees: What Rights? Which Duties'? *Res Publica* 10 (2): 135–52.

Boyd, Emily, Rachel A. James, Richard G. Jones, Hannah R. Young, and Friederike E. L. Otto. 2017. 'A Typology of Loss and Damage Perspectives'. *Nature Climate Change* 7 (10): 723–29.

Butt, Daniel. 2009. *Rectifying International Injustice: Principles of Compensation and Restitution between Nations*. Oxford; New York: Oxford University Press.
Caney, Simon. 2006. 'Environmental Degradation, Reparations, and the Moral Significance of History'. *Journal of Social Philosophy* 37 (3): 464–82.
———. 2009. 'Climate Change, Human Rights and Moral Thresholds'. In *Human Rights and Climate Change*, edited by Stephen Humphreys. Cambridge; New York: Cambridge University Press.
Dow, Kristin, Frans Berkhout, Berjamin Preston, Richard Klein, Guy Midgley, and Rebecca Shaw. 2013. 'Limits to Adaptation'. *Nature Climate Change* 3: 305–07.
Frankhauser, Samuel, Simon Dietz, and Phillip Gradwell. 2014. 'Non-Economic Losses in the Context of the UNFCCC Work Programme on Loss and Damage'. In *Policy Paper*. Centre for Climate Change Economics and Policy; London: Grantham Research Institute on Climate Change and the Environment.
García-Portela, Laura. 2020. 'Moral Responsibility for Loss and Damage: A Response to the Excusable Ignorance Objection'. *Teorema* 39 (1): 7–24.
———. 2023. 'Backward-Looking Principles of Climate Justice: The Unjustified Move from the Polluter Pays Principle to the Beneficiary Pays Principle'. *Res Publica* 29, 367–84.
Heyward, Clare. 2010. *Environment and Cultural Identity: Towards a New Dimension of Climate Justice*. Dissertation. Oxford University.
———. 2012. 'Climate Change as Cultural Injustice'. In *New Waves in Global Justice*, edited by Tom Brooks. London: Palgrave Macmillan.
Heyward, Clare, and Jörgen Ödalen. 2016. 'A Free Movement Passport for the Territorially Dispossessed'. In *Climate Justice in a Non-Ideal World*, edited by Clare Heyward and Dominic Roser. Oxford: Oxford University Press.
Huq, Saleemul, Erin Roberts, and Adrian Fenton. 2013. 'Loss and Damage'. *Nature Climate Change* 3: 947.
ICC, Inuit Circumpolar Conference. 2005. 'Petition To The Inter-American Commission on Human Rights Seeking Relief From Violations Resulting from Global Warming Caused By Acts and Omissions of the United States'. https://climatecasechart.com/non-us-case/petition-to-the-inter-american-commission-on-human-rights-seeking-relief-from-violations-resulting-from-global-warming-caused-by-acts-and-omissions-of-the-united-states/
Jaggar, Alison M. 2006. 'Reasoning About Well-Being: Nussbaum's Methods of Justifying the Capabilities'. *Journal of Political Philosophy* 14 (3): 301–22.
Kreft, S., S. Harmeling, and K. Warner. 2012. 'Framing the Loss and Damage Debate'. In *A Conversation Starter by the Loss and Damage in Vulnerable Countries Initiative*. Berlin: Germanwatch.
Kreienkamp, Julia, and Lisa Vanhala. 2017. 'Climate Change Loss and Damage'. In *Policy Brief*. Brussels: Global Governance Institute.
Linnerooth-Bayer, JoAnne, Swenja Surminski, Laurens M. Bouwer, Ilan Noy, and Reinhard Mechler. 2019. 'Insurance as a Response to Loss and Damage'? In *Loss and Damage from Climate Change: Concepts, Methods and Policy Options*, edited by Reinhard Mechler, Laurens M. Bouwer, Thomas Schinko, Swenja Surminski, and JoAnne Linnerooth-Bayer, 483–512. Climate Risk Management, Policy and Governance. Cham: Springer International Publishing. https://doi.org/10.1007/978-3-319-72026-5_21.
Mechler, R., C. Singh, K. Ebi, R. Djalante, A. Thomas, R. James, P. Tschakert, M. Wewerinke-Singh, T. Schinko, D. Ley, J. Nalau, L. M. Bouwer, C. Huggel, S. Huq, J. Linnerooth-Bayer, S. Surminski, P. Pinho, R. Jones, E. Boyd, and A. Revi. 2020. 'Loss and Damage and Limits to Adaptation: Recent IPCC Insights and Implications for Climate Science and Policy'. *Sustainability Science* 15 (4): 1245–51. https://doi.org/10.1007/s11625-020-00807-9.

Mechler, Reinhard, and Thomas Schinko. 2016. 'Identifying the Policy Space for Climate Loss and Damage'. *Science* 354 (6310): 290–92.

Meyer, Lukas. 2015. 'Intergenerational Justice'. In *The Stanford Encyclopedia of Philosophy*, edited by Edward N. Zalta. https://plato.stanford.edu/archives/sum2021/entries/justice-intergenerational/.

Meyer, Lukas, and Dominic Roser. 2009. 'Enough for the Future'. In *Intergenerational Justice*, edited by Lukas Meyer and Axel Gosseries. Oxford: Oxford University.

Morrissey, James, and Anthony Oliver-Smith. 2013. 'Perspectives on Non-Economic Loss and Damage'. *Loss and Damage in Vulnerable Country Initiative*. https://uploads-ssl.webflow.com/605869242b20501f9a579e7a/628f8033a196affe0f08d1e6_Perspectives_on_NELD.pdf

Nussbaum, Martha. 1997. 'Capabilities and Human Rights'. *Fordham Law Review* 66 (2): 273.

———. 2007. *Frontiers of Justice: Disability, Nationality, Species Membership*. Cambridge: Harvard University Press.

———. 2011. *Creating Capabilities*. Cambridge: Harvard University Press.

Otto, Friederike. 2018. *Angry Weather*. London: Greystone books. https://greystonebooks.com/products/angry-weather.

Page, Edward A. 2008. 'Distributing the Burdens of Climate Change'. *Environmental Politics* 17 (4): 556–75.

Page, Edward A., and Clare Heyward. 2016. 'Compensating for Climate Change Loss and Damage'. *Political Studies* 65 (2): 356–72.

Parfit, Derek. 1984. *Reasons and Persons*. Report. Oxford: Clarendon.

Preston, Christopher J. 2017. 'Challenges and Opportunities for Understanding Non-Economic Loss and Damage'. *Ethics, Policy & Environment* 20 (2): 143–55.

Puig, Daniel. 2022. 'Re-Conceptualising Climate Change-Driven Loss and Damage'. *International Journal of Global Warming* 27 (2): 202–212.

Schinko, T., R. Mechler, and S. Hochrainer-Stigler. 2018. 'The Risk and Policy Space for Loss and Damage: Integrating Notions of Distributive and Compensatory Justice with Comprehensive Climate Risk Management'. In *Loss and Damage from Climate Change. Concepts, Methods and Policy Options*, 83–110, edited by R. Mechler, L. Bouwer, T. Schinko, S. Surminski, and J. Linnerooth-Bayer. Cham: Springer.

Sen, Amartya. 1999. *Development as Freedom*. Oxford; New York: Oxford University Press.

———. 2009. *The Idea of Justice*. Cambridge: Harvard University Press.

Serdeczny, Olivia Maria, Steffen Bauer, and Saleemul Huq. 2018. 'Non-Economic Losses from Climate Change: Opportunities for Policy-Oriented Research'. *Climate and Development* 10 (2): 97–101.

Surminski, Swenja, and Ana Lopez. 2014. 'Concept of Loss and Damage of Climate Change – a New Challenge for Climate Decision-Making? A Climate Science Perspective'. *Climate and Development* 7 (3): 267–77. https://doi.org/10.1080/17565529.2014.934770.

UNFCCC. 2012. *A Literature Review on the Topics in the Context of Thematic Area 2 of the Work Programme on Loss and Damage: A Range of Approaches to Address Loss and Damage Associated with the Adverse Effects of Climate Change. FCCC/SBI/2012/INF.14.*

———. 2013. *Non-Economic Losses in the Context of the Work Programme on Loss and Damage*. http://UNFCCC,.int/resource/docs/2013/tp/02.pdf.

———. 2022. 'Report of the Conference of the Parties on Its Twenty-Seventh Session, Held in Sharm El-Sheikh from 6 to 20 November 2022'. *FCCC/CP/2022/10.*

Wallimann-Helmer, Ivo. 2015. 'Justice for Climate Loss and Damage'. *Climatic Change* 133 (3): 469–80.

Wallimann-Helmer, Ivo, Lukas Meyer, Kian Mintz-Woo, Thomas Schinko, and Olivia Serdeczny. 2018. 'The Ethical Challenges in the Context of Climate Loss and Damage'. In

Loss and Damage from Climate Change. Concepts, Methods and Policy Options, edited by R Mechler, I. Bouwer, T. Schinko, S. Surminski, and J. Linnerooth-Bayer. Cham, Switzerland: Springer.

Warner, Koko, and Kees Van der Geest. 2013. 'Loss and Damage from Climate Change: Local-Level Evidence from Nine Vulnerable Countries'. *International Journal of Global Warming* 5 (4): 367.

3 In search for a justified rectificatory justice principle

Historical considerations are at the core of rectificatory justice. Achieving rectificatory justice requires repairing an injustice while acknowledging and giving normative significance to its sources, such as the actions that lead to that injustice. How could we achieve, then, rectificatory climate justice thusly conceived?

We know that climate change has been caused by emission-generating activities throughout history and very intensively since the period of industrialization. Climate ethicists have proposed to acknowledge the source of climate injustice by targeting *those closely connected* to emission-generating activities as the duty-bearers of rectification, namely, polluters and beneficiaries of pollution. These two forms of addressing rectificatory justice have taken the form of two different justice principles: the Polluter Pays Principle (PPP) and the Beneficiary Pays Principle (BPP).[1]

In a nutshell, the PPP states that polluters should bear the burdens associated with emission-generating activities because they are responsible for their adverse effects. This principle follows the maxim that if 'you broke it, you fix it' (Caney 2006, 2010) or 'clean up your own mess' (Shue 1999; Gardiner 2016). The BPP states that those benefitting from climate change–inducing activities should bear the burdens associated with tackling the adverse effects of climate change. The BPP emerged in climate justice literature as an alternative to the PPP based on certain objections pressed against this first principle. Often, defenders of the BPP have claimed that this principle preserves the backward-looking elements of the PPP, but it solves the practical and theoretical challenges associated with the first principle. In this way, the BPP is taken to provide a better account of our backward-looking intuitions concerning climate justice.

Very often, defenders of the BPP have assumed that their principle is free from the problems associated with the PPP. In this chapter, I focus on two objections: the Causation Objection and the Excusable Ignorance Objection.[2] Various scholars have argued that the BPP does not encounter at least one of these objections, which provides a reason to prefer the BPP over the PPP (Gosseries 2004; Meyer and Roser 2010; Caney 2010; Bell 2011; Page 2012; Meyer 2013; Baatz 2013; Duus-Otterström 2014; Heyd 2017). Here, I challenge that underlying assumption. I argue that shifting from the PPP to the BPP based on either of these objections might be unjustified because the BPP might be affected either by the same objections or by the same considerations that gave rise to the objections.

DOI: 10.4324/9781003399889-3

Those considerations do not necessarily give us reason to prefer the PPP over the BPP either. However, they constitute a reason to further explore the possible circumvention to their objections, thereby providing a satisfactory account of rectificatory climate justice. In this book, I have chosen to show how a rectificatory account of climate justice can be grounded on a defense of the PPP (a task that will be undertaken in the next chapter). Nonetheless, this does not preclude the possibility of providing an account of rectificatory justice based on the BPP. This is simply not the option I have chosen.

3.1 Two objections against the polluter pays principle

The Causation Objection states that, without clear knowledge about which specific weather events are caused by emission-generating activities, polluters cannot be made to pay for the negative effects associated with those weather events, or EWEs (Adler 2007; Caney 2010; Farber 2017; Wallimann-Helmer et al. 2018). The reason is that we simply might not know whether those EWEs have occurred or will occur due to polluting activities at all. Climate scientists are confident that climate change has been caused by greenhouse gas (GHG) emissions. They are also highly confident that slow-onset events such as sea-level rises are caused by climate change (IPCC 2014). However, it is more difficult to know whether particular EWEs (such as extreme temperatures, heatwaves, droughts, floods, extreme rainfall, etc.) have been or will be caused by anthropogenic influences on the climate system, because EWEs can also happen due to natural variability in a world without climate change. Although detection and attribution studies have improved significantly in the last decade (Stuart-Smith et al. 2021), they are still in development and not free of important objections and problems (Huggel et al. 2015; Shepherd 2014).

Note that this objection affects particularly the application of the PPP to adaptation and L&D duties but not to mitigation duties. Mitigation duties might be required because we know that climate change will likely cause harm in the future and that additional emissions are likely to contribute to that harm, although we do not know exactly *where* that harm will occur.[3] Nothing of that requires connecting emission-generating activities and with the local manifestations of climate change. But adaptation duties (understood as *climate change–related* duties) require knowing which regions are projected to be affected by (at least some) weather events that have climate change as their main driver. Thus, if adaptation duties ought to be distributed to polluters, a connection between emissions and the location of those foreseeable harmful impacts needs to be established. Otherwise, polluters could rightfully ask why their money is used to tackle foreseeable harm without knowing whether that potential harm is connected to their emissions. Likewise, those seeking L&D measures for the harmful impacts of climate change based on the PPP need to show a connection between polluters' emissions and the harm they suffered.[4]

The Excusable Ignorance Objection states that no one should be held accountable for the effects of their actions if these were unknown and could not reasonably have been foreseen. Arguably, this is the case for historical emissions. Before the publication of the First IPCC Report in 1990, the scientific community had not yet reached consensus about the adverse effects of GHG emissions on the climate system. Governments and citizens alike could not be expected to know about the adverse effects of emission-generating activities. Many scholars have considered this objection to be one of the most powerful against the application of the PPP for pre-1990 emissions (Gosseries 2004; Caney 2006, 2010; Bell 2011; Page 2008, 2012; Meyer and Roser 2010; Wündisch 2017). This objection has also been relevant in climate negotiations (Gardiner 2016, 111). Admittedly, the objection is relatively limited because it only affects emissions emitted until roughly 1990. However, and importantly, these historical emissions still represent around half of total global emissions.[5]

Unless we want to challenge the empirical assumptions concerning the availability of information about the negative effects of climate change and the role of emission-generating activities, the only way to apply the PPP to historical emissions is as a principle of strict liability. That is, the PPP needs to be understood as a principle that allocates to an agent the duty to deal with the harm associated with her action, phi, 'irrespective of any steps that she took in order not to phi and irrespective of whether she knew or had reason to know that she was phi-ing [including any steps she took to find out whether she was about to phi]' (Gardner 2011, 207). This understanding of the principle departs from one that relies on attributions of culpability and that seeks to punish emitters, at least for historical emissions. In the following, for the sake of the argument, I propose to accept the empirical assumptions behind the Excusable Ignorance Objection and, thus, to accept that the PPP can only work as a principle of strict liability when applied to historical emissions. Moreover, my discussion will be based on the objections that the PPP faces as a principle of strict liability.[6]

These objections have led some scholars to propose an alternative principle, the BPP, at least to cover duties related to a part of overall emissions. In the next section, I introduce this principle in more detail and show how, at first glance, it could presumably surmount these objections.

3.2 The beneficiary pays principle and some intuitive reactions to the objections

A more general version of the BPP states that beneficiaries of an injustice should bear the burdens associated with the injustice. With climate change, the principle is generally understood as stating that those benefitting from emission-generating activities should bear the burdens associated with tackling climate injustice or the harmful effects of climate change.[7] In the context of climate change and historical emissions, the principle is most widely understood as possessing the three salient following characteristics.

First, the BPP usually assumes that the beneficiaries are innocent. This means that, in receiving their benefits, they did not do anything for which they can be considered culpable, such as inducing or participating in unjust or potentially harmful actions. Consequently, it is not their culpability that grounds their duties. Second, the principle does not require that the original perpetrators (i.e., emitters) are culpable for their unjust or harmful actions. Thus, the principle applies when the emitters from whom beneficiaries receive their benefits did not know about the negative consequences of climate change or when they could have not avoided engaging in climate change–inducing activities. This characteristic allows the principle to account for historical emissions, and thus it supports the most overarching version of the BPP when applied to climate change.[8]

Third, the principle is usually understood to be victim centered in two ways: (i) The principle is motivated by a concern for the actual or potential victims and it seeks to prevent or alleviate their suffering (Baatz 2013, 2016; Lawford-Smith 2014; Couto 2018). (ii) The principle is also victim centered in a more specific way, because it is not only focused on solving the situation of victims of injustices or undeserved harm in general, but also focused on the suffering of those victims that are likely to be affected by the activities from which the beneficiaries, and thus potential duty-bearers, benefit. This feature is grounded in the common source of benefits and disadvantages, which often operates as a justification of the principle (Duus-Otterström 2017). As Page has put it: 'profiting from activities that impose climatic disadvantages . . . here, triggers a remedial duty on the part of the beneficiaries . . . solely because the disadvantages and benefits share *common origins*' (Page 2012, 313).[9] In the same vein, others have argued that this is a principle that seeks to even out the benefits and potential harm associated with climate change (Meyer and Roser 2010; Baatz 2013; Meyer 2013).

This victim-centeredness characterizes the version of the BPP that I address here. Alternative versions remain excluded from my discussion. Nonetheless, I believe that this feature makes this principle a genuine principle of climate justice in comparison to other versions of the BPP.[10] Take, for instance, a purely beneficiary-centered version of the BPP. The rationale of this principle is that it is wrong in itself for the beneficiary to keep certain benefits associated with injustices or undeserved harm because they are tainted, and the point of this principle is to require beneficiaries to just surrender their benefits (Couto 2018, 2172). Thusly defined, this principle can hardly work as a proper principle of climate justice because, arguably, a principle of climate justice requires not only removing resources from certain agents but also using them to solve certain problems caused by climate change. Or take, for instance, an 'undirected disgorgement BPP' (Duus-Otterström 2017; similar Goodin 2013). According to this principle, tainted benefits acquired from unjust or harmful activities should not only be given up, as with a purely beneficiary-centered principle, but should also be added to society's general pool of resources and eventually used to solve injustices or alleviate undeserved harm. Nonetheless, this principle cannot work as a proper principle of climate justice when thus described because a principle of climate justice arguably requires certain resources to be used to address the actual or potential undeserved harm associated with climate change. Neither principle can properly work as a principle specifically of climate justice: a principle that seeks to tackle climate injustice, the harmful effects of climate change, and other problems associated with climate change.

The BPP has often been described as a hybrid principle that combines backward-looking aspects with forward-looking ones and that has the advantages of both without their disadvantages (Page 2012). Forward-looking principles consider the reasons we have today to make the world a better place now and usually take into account individuals' abilities to bear these burdens. This BPP relies partly on a forward-looking rationale because it is focused on alleviating underserved harm and relies on agents' ability to solve the problem with the benefits they received. However, it is also backward-looking because it does not ignore the sources of the problem and 'isolates for redistribution only those benefits that are strongly connected to *climate change producing acts*'[11] (Page 2012, 313).

Advocates of the BPP have proposed this principle as an alternative to the PPP. Many climate justice scholars have moved, in one way or the other, from the PPP to the BPP in response to at least one of the objections introduced in the previous section (see Gosseries 2004; Meyer and Roser 2010; Caney 2010; Bell 2011; Page 2012; Meyer 2013; Baatz 2013, 2016; Duus-Otterström 2014; Heyd 2017).[12] Interestingly, however, defenders of the BPP do not address how this principle can circumvent these objections. Instead, they only assume that it can do so.

Admittedly, there might be reasons to believe that the BPP is not open to these objections. First, concerning the Causation Objection, this principle does not focus on the link between emission-generating activities and their effects on certain geographical locations. Second, it seems to be unaffected by the Excusable Ignorance Objection because the grounds for making beneficiaries bear the burdens associated with climate change are the resources that they might have enjoyed, not the actions that caused climate change. These reasons could explain the assumption that the BPP is not open to the objections pressed against the PPP. However, I believe that a closer examination shows that the BPP is indeed vulnerable at least to the concerns underlying these objections.

3.3 The causation objection and the beneficiary pays principle

To see whether the BPP also relies on the type of connection required by the Causation Objection, we need to recapitulate and explore the principle a little more. The principle requires beneficiaries of emission-generating activities to bear the burdens associated with addressing the adverse effects of climate change. The principle 'isolates for redistribution only those benefits that are strongly connected to *climate change producing acts*' (Page 2012, 313).[13] That is, the benefits that are up for redistribution are those coming from emission-generating activities. This reference to the link between benefits and emission-generating activities appears repeatedly in the literature on the BPP. Similarly, Meyer (2013) has highlighted that 'the goods in question [whose redistribution is called for] are the benefits that people realize in *carrying out actions that unavoidably have emissions as a side-product*' (Meyer 2013, 600).[14] The underlying assumption in the current discussion is that beneficiaries should bear the duties associated with tackling climate change because they have benefited from the activities causing the problem, which are emission-generating activities (see also Caney 2010; Baatz 2013; Schüssler 2011; Karnein 2017).

Conversely, we have the burdens that are to be paid for with these benefits. As we saw earlier, the relevant burdens are those associated with undertaking or paying for adaptation and L&D. Moreover, as we saw before, the 'common source' justification for the BPP excludes beneficiaries of emission-generating activities from the duty to bear burdens associated with injustices or undeserved harm unconnected to GHG emissions, such as burdens associated with addressing the negative effects of unjust wars, terrorist attacks, sexual assaults, gender violence, and rare diseases. Most importantly, they also exclude the injustices or undeserved harm caused by environmental impacts resulting from natural variability. None of this (potential) harm is derived from emission-generating activities, and therefore, it is not the duty of these beneficiaries to bear the burdens of preventing or alleviating it, according to this principle. Thus, attributing duties to alleviate climate change–related harm to beneficiaries of emission-generating activities requires discriminating between weather events caused by anthropogenic forcing and those caused by natural variability.

These remarks show that the BPP runs into the Causation Objection for both L&D and adaptation burdens. Let us start with L&D burdens, for which the application of the Causation Objection is perhaps clearer. According to the BPP, those benefitting from GHG emissions are the ones who should bear the burdens of L&D because those emissions both benefitted them and caused harm. The rationale behind the principle is that beneficiaries of emissions should respond to the harm caused by the emissions because benefits and harm share the same source: the emissions. But this does not seem to sidestep the problem of causation, because the GHG emissions would still need to be shown to be involved in the causation of the harm for which L&D is required. The connection between emissions and the harm still needs to be proven. If causation were not proven but we used the benefits of GHG emissions to compensate those affected by environmental harmful events happening in specific geographical locations, we would run the risk of using these benefits to address the harm that does not share a common source with the benefits. But this is not what the benefits are for, according to the BPP. Therefore, applying the BPP to compensatory duties requires the demonstration of a causal link between GHG emissions and their effects. *As long as* we think this is a difficult or even an impossible task, the BPP runs into the Causation Objection.

Is this any different for adaptation to the foreseeable harmful effects of anthropogenic climate change? I do not think so. According to the BPP, those benefitting from GHG emissions are the ones who should bear the burdens of adaptation because those emissions both benefitted them and will foreseeably cause harm to third parties if adaptation measures are not undertaken. The rationale behind the principle is that beneficiaries of emissions should avoid the potential harm that would be caused by emissions because their benefits and that potential harm share the same source: the emissions. Again, this does not seem to sidestep the problem of causation. The GHG emissions would still need to be shown to be significantly involved in the causation of the foreseeable harm for which adaptation is required. If causation were not proven but we used the benefits of GHG emissions for adaptation measures to avoid the foreseeable harmful events happening in specific geographical locations due to weather events, we would run the risk of using these benefits to address the potential harm that does not share a common source with the benefits. Remember: those weather events might still be caused by changing environmental conditions that are not due to climate change. But the BPP is not meant to distribute burdens for avoiding the foreseeable harmful impacts of weather events that are not due to climate change, as it is not meant to distribute burdens concerning the foreseeable harmful effects of rare diseases or terrorist attacks. Arguably, it would be unfair to ask beneficiaries of emissions to bear the burdens of adaptation to ACC if we don't even know whether those foreseeable harmful effects will be caused by anthropogenic climate change (i.e., emission-generating activities) at all. Therefore, applying the BPP to adaptation duties requires the demonstration of a causal link between GHG emissions and potential harm. *As long as* we think this is a difficult or even an impossible task, the BPP runs into the Causation Objection also when applied to adaptation duties.

3.4 The excusable ignorance objection and fairness considerations

The Excusable Ignorance Objection states that no one should be held accountable for the effects of their actions if these were unknown and could not reasonably have been foreseen. However, one might wonder why this is the case. Climate ethicists have rarely provided a deeper justification for the Excusable Ignorance Objection: they rarely explain what the problem is with holding someone accountable for the effects of their actions when these were undertaken under circumstances of excusable ignorance. Instead, they merely state that excusable ignorance is a problem when trying to hold agents liable for historical emissions.

Various reasons might be advanced to reject holding people accountable for the effects of their actions if these were unknown and could not reasonably have been foreseen. However, an exhaustive analysis is beyond the scope of this book. Instead, I would like to propose and explore one possible interpretation of the concerns motivating the Excusable Ignorance Objection. I argue that the Excusable Ignorance Objection might be motivated by fairness considerations and that these considerations apply not only to the PPP but also to the BPP. Hence, if I am right, this conclusion precludes moving from the PPP to the BPP in response to the Excusable Ignorance Objection.[15]

3.4.1 A fresh look at the excusable ignorance objection

In this section, I offer a plausible interpretation of the motivation behind the Excusable Ignorance Objection in two steps. First, I show that the concern behind the Excusable Ignorance Objection might not be best expressed by saying that people should not be held accountable for the effects of their actions if these were unknown and could not reasonably have been foreseen. Instead, the concern might be that people should not be required to bear burdens they could not have expected being required to bear. I support my interpretation by showing how retrospective and prospective principles of strict liability are unequally affected by the original formulation of the Excusable Ignorance Objection. Second, I argue that this more precise formulation of the Excusable Ignorance Objection might be supported by fairness considerations: making people bear these burdens is unfair because it frustrates their legitimate expectations and truncates their ability to plan and execute their mid- and long-term life plans.

To begin, we first need to clarify the kind of principle against which the Excusable Ignorance Objection is usually raised in climate justice debates. As we saw, this objection affects the PPP applied to pre-1990 emissions. The PPP works here as a principle of strict liability. Such a principle imposes liability for the harmful effects of actions

> irrespective of any steps that she took in order not to phi and irrespective of whether she knew or had reason to know that she was phi-ing [including any steps she took to find out whether she was about to phi].
>
> (Gardner 2011, 207)

We should differentiate between prospective and retroactive strict liability.[16] Prospective strict liability, or strict legal liability, is applied when principles of strict liability already exist as part of some regulatory scheme or legal system and their justification lies in the future-orientated implications of these principles. This understanding of strict liability is common in regulations concerning environmental pollution.[17] For instance, a strict liability principle may state that if a chemical company pollutes a river, the company will be held liable for the environmental damage caused by the pollution regardless of whether the company made all reasonable efforts to avoid this damage and regardless of whether the company had good reasons to believe that these would be the results of its action. This feature makes this a principle of strict liability, but its justification is forward-looking. The existence and application of such a principle are justified by the general and prospective beneficial consequences of holding companies, people, and other entities liable for the harmful effects of their actions when they manipulate potentially harmful substances. In such circumstances, a principle of strict liability may be justified as a way of distributing the risks and costs of damages effectively. It may well be that making those who cause harm pay for the adverse effects of their actions is the best way to ensure that people take reasonable steps to avoid harm or even that it is the best way to ensure that costs do not fall on victims.

Note that this prospective justification requires that everyone must have the opportunity to be aware of it: information about the applicability of strict liability must be disseminated ahead of time in a way that any person can have reasonable access to it. Further, agents must have a choice about whether and how to participate in the activities regulated by these principles of strict liability (Wündisch 2017, 845). Arguably, if these principles are to disincentivize certain activities and make people maximally careful, they need to be widely known in advance. Thus, people are aware of the costs they might incur by engaging in certain activities and can decide whether and how to engage in these activities. Otherwise, the disincentivizing mechanism will not work.

Usually, proponents of the Excusable Ignorance Objection do not object to this application of strict liability. They even agree that these principles are a good way of distributing the risks associated with certain potentially harmful substances. A system of strict liability, they acknowledge, may establish the right 'incentives either to refrain from especially dangerous activity altogether or to be exceedingly careful when engaging in it' (Moellendorf 2014; similarly Bell 2011 and Wündisch 2017). They believe that the Excusable Ignorance Objection does not apply in these circumstances because, although those causing harm may not have been able to know about the negative effects of their actions, they were informed beforehand of the burdens they would need to bear in case of accidents, and yet they decided nonetheless to undertake these actions.

The situation is different for retrospective strict liability. Strict liability is applied retrospectively when no preexisting legal scheme assigns liability for the negative effects of certain actions. Agents are held liable for the effects of their actions regardless of whether they made all reasonable efforts to avoid these effects and regardless of whether they had good reason to believe that the effects would result from their actions. However, unlike prospective strict liability, people are not informed beforehand of the burdens they need to bear for the possible negative effects of their actions. Instead, the causal connection between agents' actions and their negative effects is taken to be morally relevant to attributing liability for the costs of the negative effects.

To summarize, prospective and retrospective liability have one thing in common. In both cases, people might be held liable for the negative effects of their actions even if they were unknown to them and could not reasonably have been foreseen. However, they differ in that with prospective liability, people are only held liable for the negative effects of their actions if they were informed beforehand of the burdens they would be required to bear if some accident occurs, whereas in retrospective liability, people are held liable for the consequences of their actions even if they were not made aware of the burdens they would be required to bear were their actions to trigger negative effects.

Proponents of the Excusable Ignorance Objection accept the prospective application of strict liability principles but not the retrospective application. That is, they do not take issue with all the principles that attribute liability for actions whose negative effects could have not been foreseen, such as principles of prospective strict liability. However, they do take issue with the application of principles that make agents bear burdens they could not have expected being required to bear. For this reason, it seems that the concern behind the Excusable Ignorance Objection is about making agents bear burdens they could not have expected being required to bear. This rationale applies to historical emissions because if people did not know and could not reasonably have known about the negative effects of their actions, they could not expect being required to bear the burdens associated with these negative effects.

It is plausible to assume that this concern might be grounded in fairness considerations. Arguably, it is unfair to make people bear burdens they could not have expected being required to bear because this undermines their ability to act as rational planners and executors of plans. Existing laws and regulations, or, in Rawlsian terms, the rules of the basic structure, provide the background for people's legitimate expectations (Rawls 1971). This stable and relatively permanent framework of expectations provides in turn the basis on which rational planners can consistently and effectively pursue their own ends (Buchanan 1975, 422). That is, when people plan and execute their life plans, they do so against a background of legitimate expectations provided by current regulations and law. These include, among others, expectations about what they are permitted to do and what the consequences of their actions would be if they act against current regulations and laws, including what kind of burdens they might be required to bear in such circumstances. It is on this basis that people develop their mid- and long-term plans, which constitute a solid basis for them living a good life (Rawls 1971, 497–16; Williams 1973, 116–17).

However, if people are required to bear unexpected burdens, this frustrates their legitimate expectations and undermines their ability to pursue their mid- and long-term life plans, which in turn has pernicious effects on their ability to live a good life. For instance, bearing unexpected burdens undermines the background of overall resources on which they rely to go about their lives. Hence, imposing unexpected burdens on people harms them in morally significant ways, undermining their ability to pursue their life plans according to their own conception of the good life (Meyer and Sanklecha 2011, 2014).

According to the interpretation I have offered here, the problem that some climate scholars might see with the retroactive application of strict liability with the PPP can be understood to emerge from these fairness considerations. The motivation behind the Excusable Ignorance Objection might be that people should not be required to bear the burdens associated with tackling the negative impacts of climate change because they could not have expected being required to bear these burdens and that imposing these burdens on them would be unfair because it would ultimately undermine their ability to execute their mid and long-term plans. Thus, the Excusable Ignorance Objection might be grounded in a fairness concern that applies where people did not have the relevant information and could not be expected to have had it, either about the relevant moral facts or about the possibility of being held liable for certain costs.

3.4.2 *Fairness considerations, the beneficiary pays principle, and replies to some objections*

In the previous section, I have argued that the Excusable Ignorance Objection might be grounded in fairness considerations concerning the harmful impacts of the imposition of unexpected burdens on people that affect their planning and execution of life plans. In this subsection, I propose to assess how this concern may also appear in the application of the BPP.

When individuals pursue certain life plans, they do so against a background of resources that are available to them now and will be in the future. In industrialized countries, this background is heavily influenced by the benefits accumulated over centuries from emission-generating activities (Meyer and Sanklecha 2011, 2014). Industrialized countries and their populations have long relied on the use of these resources. The longer they have been relying on these resources, the stronger the expectations are that they will be able to use them in the future, and the stronger is the importance of the availability of the resources to achieving their plans (Moore 2017). Benefits acquired through emission-generating activities constitute the background conditions against which rational beings plan and execute their life plans in industrialized countries.

The BPP affirms that beneficiaries should devote these resources to tackling climate change. This means that these benefits will be taken from those who have been relying on them for a long time and that the life plans and expectations depending on the use of these resources will therefore be frustrated. Note that this leaves us with the same problem that we faced with the PPP when applied as a retroactive strict liability principle. In both cases, people are asked to bear burdens they could not have expected being required to bear, which has important moral implications for the development of their mid- and long-term life plans. Thus, the same fairness concern appears again and affects the BPP.

Admittedly, the force of this objection depends on how long beneficiaries have enjoyed these benefits before knowing about their connection to the injustice or undeserved harm, because that will determine how many of their life plans rely on the use of the benefits and the extent to which they are unexpectedly frustrated. With climate change, the benefits that people enjoyed before learning of the harm attached to them are huge. Very likely, most of the basic infrastructure of highly industrialized countries was developed before people learned about the negative effects of climate change. This also explains why the BPP significantly frustrates people's life plans: because long-lasting benefits have also solidified expectations about long-held mid-term and future plans.

Defenders of the BPP might object that the frustration of people's legitimate expectations is not as worrisome as the previous argument suggests and even that it is not necessarily unfair to frustrate their expectations. Perhaps these expectations should not be so strongly protected. Hence, the fact that beneficiaries are burdened should not be a reason to refrain from making them bear the costs associated with addressing climate change. The BPP's defenders might argue that if we share the concern, as we should, that those negatively affected by climate change should not bear the associated burdens, someone else should. In this sense, being burdened is simply the consequence of having to deal with certain justice claims. These burdens might be justified, for instance, because they help in minimizing injustices. This is what Alexandra Couto has called the minimizing injustice argument in support of the BPP (2018, 2179). One might even argue analogously that wealthy people will also be burdened when new taxes are imposed on them to tackle unjust distribution in a society. But the burdens created by taxing wealthy people might not trigger fairness concerns when we weigh these burdens against other distributive justice considerations, such as that arising from the regrettable situation of victims of injustice or undeserved harm.

Against this objection, we should note that my argument does not even need to rely on the idea that frustrating people's legitimate expectations is worrisome, unfair, or even something we should avoid. Instead, remember that my argument is conditional in nature. My point is that *if* the PPP is affected by the Excusable Ignorance Objection and the motivation behind the Excusable Ignorance Objection is based on these fairness considerations, *then* the same fairness considerations apply to the BPP in the context of climate change. Hence, my initial argument still holds: at least some concerns behind the Excusable Ignorance Objection also affect the BPP and to the same extent as the PPP, which undermines moving from the PPP to the BPP based on these grounds.

Admittedly, one might believe that frustrating legitimate expectations is not unfair under these circumstances and that minimizing the suffering of victims should take priority over these considerations. However, in the absence of further explanations, this minimizing injustice argument cannot on its own provide specific support only to the BPP. If what matters is to avoid a situation in which the victim alone bears the burdens, other principles could also be said to achieve this (Couto 2018, 2180). These include principles that distribute remedial duties according to whom has the highest capacity or just randomly, and even more interestingly, principles of strict liability such as the PPP itself. The point here is not only that the minimizing injustice argument cannot provide specific support only to the BPP but that such an argument can also provide support to the PPP. Thus, again, such an argument cannot provide a reason to support the BPP that cannot be equally applied to the PPP.

3.5 Conclusion

Various climate ethicists have proposed moving from the PPP to the BPP as a rectificatory and backward-looking principle of climate justice based on the Causation Objection or the Excusable Ignorance Objection. Implicitly, they have assumed that the BPP is not vulnerable to these objections. In this chapter, I have challenged that assumption and argued that moving from the PPP to the BPP in response to any of these objections might be unjustified because the BPP may be affected by the same objections or by the same considerations that give rise to these objections.

First, I have shown that the BPP is subject to the Causation Objection. In a nutshell, if the principle requires the benefits from emission-generating activities to be used to balance the harm associated with these emissions, in adaptation and L&D cases, we need to determine where the harm has occurred or will foreseeably occur. That is, we need to discriminate between environmental harmful effects that have been caused or will be caused by climate change and those that have not and will not be caused by climate change, but just by natural climate variability. According to the BPP, beneficiaries should bear the burdens associated with the former but not with the later. This requires proving the same level of causation as that involved in the Causation Objection.

Second, I have argued that the Excusable Ignorance Objection might be motivated by fairness considerations arising from the frustration of people's legitimate expectations and the truncation of their mid- and long-term life plans when they are required to bear burdens they could not have expected being required to bear. I have shown that this concern also appears in the case of the BPP because those who would be required to bear the costs of climate change through having benefitted from it have relied on these benefits for a long time. Their life plans and expectations about their future depend on the idea that they will be able to continue to use these resources, and these people will be harmed if the resources needed to address climate change–related problems are taken away from these people. Hence, if the Excusable Ignorance Objection is motivated by these fairness considerations, these considerations apply to both the PPP and the BPP. Again, this undermines the move from the PPP to the BPP based on these grounds.

Finally, I have argued that even if we believe that the frustration of legitimate expectations and mid- and long-term life plans is not a cause for very serious concern, this does not affect my main point, which is conditional in nature. The idea is that if one believes that these considerations are relevant, then one should believe they are relevant for both principles. But if one believes that they are not that relevant, then they are not relevant for either of these principles. Therefore, my main point remains untouched: one cannot justify moving from the PPP to the BPP based on this consideration.

Some might wonder where these conclusions lead. Let me lay out some of the possible implications of my analysis and thus some of the ways forward. First, if both principles face the same problems, then defenders of the BPP might want to reconsider their skepticism toward the strict liability form of the PPP. But, second, if advocates of the BPP insist on defending the purported superiority of this principle, they need to reply to these challenges or bring forward other reasons to move from the PPP to the BPP that might have not been addressed in this book. In this sense, this chapter can be either taken as a light defense of the PPP, showing that the challenges it faces do not uniquely apply to this principle, or as agenda-setting for defenders of the superiority of the BPP.

Finally, the chapter can also be taken as an agenda-setting for anyone defending a rectificatory approach to climate justice. Such an approach would need to be developed in a way that solves these two objections and perhaps others that might appear on the way. In the next chapter, I provide such an account based on the PPP.

Notes

1 Arguably, the BPP has been conceived as a hybrid principle that combines both backward- and forward-looking intuitions (see Section 3.3).

2 Other objections include the nonidentity problem (Kumar 2003; Duus-Otterström 2014) and the dead-polluters objection (García-Portela 2019; Francis 2020). Whether and to what extent any of these objections constitute reasons to move from the PPP to the BPP is not discussed here.

3 Even though we might not be fully sure that certain emissions did or will cause harm, it might simply be extremely likely that they did or will do so. For a discussion about difference-making of additional emissions, see Kagan (2011) and Broome (2019).

4 The Causation Objection, as it is described here, is related but independent from other causation concerns that might arise in the context of responsibility for climate change-related harm. An additional worry might be that it is so far very difficult to know whether an agent's emissions have caused or contributed to cause some specific harmful weather events. Unlike in my Causation Objection, the skepticism here is not about the connection between emissions generally and local impacts, but rather about *someone's* emissions and local impacts. Arguably, this can be a challenge for distributing causal and thus moral responsibility to individual agents. However, this worry might less concerning. Arguably, once one can link specific weather events with anthropogenic climate change, one could adopt a proportional division of responsibility even if proportional causal responsibility cannot be proven (see, for instance, Wündisch 2019; Harrington and Otto 2019). Quite obviously, notice that this objection would not apply to the BPP because this principle does not distribute responsibility according to emission, but to benefits. For that reason, I do not engage in this discussion here.

5 If we take 1990 as the cut-off date, historical emissions from 1750 to 1990 represent 48.6% of overall emissions. If we take 1995, historical emissions represent 55.6% of overall emissions. These refer only to CO_2 emissions. Earlier emissions, until roughly 1960, are rough estimates. Data source: Hannah Ritchie and Max Roser, 2017. 'Causation objection$_2$ and Greenhouse Gas Emissions', Published online at OurWorldInData.org (https://ourworldindata.org/co2-and-other-greenhouse-gas-emissions). Last revision: August 2020.

6 Arguably, the PPP could be understood as a principle of fault liability for later emissions. However, this does not affect the conclusions of this chapter. If the Excusable Ignorance Objection does not apply and the PPP works as a principle of fault liability, this also undermines moving from the PPP to the BPP, but I do not discuss this issue further here.

7 Those who believe that engaging in emissions-generating activities cannot be described as an injustice because of the Excusable Ignorance Objection focus on 'undeserved harms' (see Meyer and Roser 2010; Meyer 2013; Caney 2006). But others believe that regardless of the Excusable Ignorance Objection, these activities could constitute an 'objective injustice' (Duus-Otterström 2014) if considered from a time-neutral perspective (Bell 2011; Thompson 2017). The conclusions I present here are valid for both understandings of the principle.

8 The version of the principle I present here combines both the 'unjust enrichment BPP' and the 'wrongful enrichment BPP' (Page 2012; Heyd 2017), both of which have been discussed in the literature. The first does not require that perpetrators are culpable, whereas the second requires that the actions triggering the original injustice or undeserved harm were committed under culpable conditions. Arguably, a 'wrongful enrichment BPP' could be applied in circumstances when emissions-generating activities can no longer be considered innocent. But only the 'unjust enrichment BPP' can account for historical emissions. I here propose a formulation that includes both possibilities because this constitutes a broader climate justice principle that covers both historical emissions and later possibly culpable emissions.

9 Emphasis mine.

10 This does not mean that the BPP can also be partly understood as being beneficiary centered. I argue that for the principle to work as a genuine climate justice principle, it needs to at least also be victim centered in these ways, for the reasons provided in the main text.

11 Emphasis mine.

12 Two things are worth noting. First, the Excusable Ignorance Objection has been emphasized more than the Causation Objection. However, the Causation Objection has been mentioned by Gosseries (2004, 54), Baatz (2013, 95–97, 2016), Heyd (2017, 27), and Walliman-Helmer et al. (2018, 45–46 fn.3) as an objection against the PPP but not

against the BPP. Second, the principles of modified strict liability (Caney 2010) and limited liability (Bell 2011) also include this move. These principles involve applying the PPP to identify the duty bearers but limiting the extension of their duties only to the benefits acquired through their emissions (Heyward forthcoming). This principle can be read as a version of the BPP that only applies when the beneficiaries are also involved in the harmful action. Here, I understand Caney's and Bell's positions as being affected by my argument because they move from the PPP to the BPP when determining the extension of duties.

13 Emphasis mine.

14 Ibid.

15 Admittedly, the objection might be grounded in considerations other than fairness. Here, I only show that there is a plausible interpretation of the Excusable Ignorance Objection based on fairness considerations and that these also apply to the BPP. This is enough to undermine the move from the PPP to the BPP, at least to some extent. If the Excusable Ignorance Objection were grounded in other kind of considerations, they might or might not apply to the BPP. However, to my knowledge no one has provided a deeper explanation of the Excusable Ignorance Objection in other terms, and exploring such an alternative explanation here is not possible due to space constraints. In any case, this should not be a major problem if results are read in a conditional way. My point here is that *if* the Excusable Ignorance Objection is based on fairness considerations of the sort advanced here, then they also apply to the BPP.

16 Other terms such as strict 'legal' and 'moral' liability are frequently used to express the same distinction (Wündisch 2017).

17 Examples of regulatory frameworks of environmental and toxic pollution include the Comprehensive Environmental Response, Compensation, and Liability Act (CERCLA) in the United States and the 2004 Environmental Liability Directive in Europe. For an analysis of how legal systems deal with strict liability for environmental pollution and the possibility of applying these strict liability principles to the case of climate change, see Farber (2017).

References

Adler, Matthew D. 2007. 'Corrective Justice and Liability for Global Warming'. *155 University of Pennsylvania Law Review* 1859–67.

Baatz, Christian. 2013. 'Responsibility for the Past? Some Thoughts on Compensating Those Vulnerable to Climate Change in Developing Countries'. *Ethics, Policy & Environment* 16 (1): 94–110.

———. 2016. 'Compensating Victims of Climate Change in Developing Countries Justification and Realization'. Dissertation, Kiel University.

Bell, Derek. 2011. 'Global Climate Justice, Historic Emissions, and Excusable Ignorance'. *The Monist* 94 (3): 391–411.

Broome, John. 2019. 'Against Denialism'. *The Monist* 102 (1): 110–29.

Buchanan, Allen. 1975. 'Distributive Justice and Legitimate Expectations'. *Philosophical Studies* 28 (6): 419–25. https://doi.org/10.1007/BF00372903.

Caney, Simon. 2006. 'Environmental Degradation, Reparations, and the Moral Significance of History'. *Journal of Social Philosophy* 37 (3): 464–82.

———. 2010. 'Climate Change and the Duties of the Advantaged'. *Critical Review of International Social and Political Philosophy* 13 (1): 203–28.

Couto, Alexandra. 2018. 'The Beneficiary Pays Principle and Strict Liability: Exploring the Normative Significance of Causal Relations'. *Philosophical Studies* 175 (9): 2169–89.

Duus-Otterström, Göran. 2014. 'The Problem of Past Emissions and Intergenerational Debts'. *Critical Review of International Social and Political Philosophy* 17 (4): 448–69.

———. 2017. 'Benefiting from Injustice and the Common-Source Problem'. *Ethical Theory and Moral Practice* 20 (5): 1067–81.

Farber, Daniel. 2017. 'How Legal Systems Deal with Issues of Responsibility for Past Harmful Behavior'. In *Climate Change and Historical Emissions*, edited by Lukas Meyer and Pranay Sanklecha, 80–106. Cambridge; New York: Cambridge University Press.

Francis, Blake. 2020. 'In Defense of National Climate Change Responsibility: A Reply to the Fairness Objection'. *Philosophy & Public Affairs* 49 (2).

García-Portela, Laura. 2019. 'Individual Compensatory Duties for Historical Emissions and the Dead-Polluters Objection'. *Journal of Agricultural and Environmental Ethics* 32 (4): 591–609.

Gardiner, Stephen. 2016. 'In Defense of Climate Ethics'. In *Debating Climate Ethics. Debating Ethics*, edited by Stephen M. Gardiner and David A. Weisbach. New York: Oxford University Press.

Gardner, John. 2011. 'Some Rule-of-Law Anxieties about Strict Liability in Private Law'. In *Private Law and the Rule of Law*, edited by Lisa. M Austin and Dennis Klimchuk, 207–23. London: Oxford University Press.

Goodin, Robert E. 2013. 'Disgorging the Fruits of Historical Wrongdoing'. *American Political Science Review* 107 (03): 478–91.

Gosseries, Axel. 2004. 'Historical Emissions and Free Riding'. *Ethical Perspectives* 4 (1).

Harrington, L. J., and F. E. L. Otto. (2019). 'Attributable Damage Liability in a Non-Linear Climate'. *Climatic Change* 153: 15–20.

Heyd, David. 2017. 'Climate Ethics, Affirmative Action and Unjust Enrichment'. In *Climate Change and Historical Emissions*, edited by Lukas Meyer and Pranay Sanklecha, 22–45. Cambridge; New York: Cambridge University Press.

Heyward, Clare. 2021. 'Is the BPP Redundant in Climate Justice'? *Norsk Filosofisk Tidsskrift* 56: 125–36.

Huggel, Christian, Dáithí Stone, Hajo Eicken, and Gerrit Hansen. 2015. 'Potential and Limitations of the Attribution of Climate Change Impacts for Informing Loss and Damage Discussions and Policies'. *Climatic Change* 133 (3): 453–67.

IPCC. 2014. *Climate Change 2014: Synthesis Report*, edited by R. K. Pachauri and L. A. Meyer, Core Writing Team. Geneva: Intergovernmental Panel on Climate Change.

Kagan, Shelly. 2011. 'Do I Make a Difference'? *Philosophy and Public Affairs* 39 (2): 105–41.

Karnein, Anja. 2017. 'Asking Beneficiaries to Pay for Past Pollution'? In *Climate Change and Historical Emissions*, edited by Lukas Meyer and Pranay Sanklecha. Cambridge; New York: Cambridge University Press.

Kumar, Rahul. 2003. 'Who Can Be Wronged'? *Philosophy & Public Affairs* 31 (2): 99–118.

Lawford-Smith, Holly. 2014. 'Benefiting from Failures to Address Climate Change: Benefiting from Failures to Address Climate Change'. *Journal of Applied Philosophy* 31 (4): 392–404.

Meyer, Lukas. 2013. 'Why Historical Emissions Should Count'. *Chicago Journal of International Law* 13: 597.

Meyer, Lukas, and Dominic Roser. 2010. 'Climate Justice and Historical Emissions'. *Critical Review of International Social and Political Philosophy* 13 (1): 229–53.

Meyer, Lukas, and Pranay Sanklecha. 2011. 'Individual Expectations and Climate Justice'. *Analyse & Kritik* 33 (2): 449–72.

———. 2014. 'How Legitimate Expectations Matter in Climate Justice'. *Politics, Philosophy & Economics* 13 (4): 369–93.

Moellendorf, Darrel. 2014. *The Moral Challenge of Dangerous Climate Change: Values, Poverty, and Policy*. New York: Cambridge University Press.

Moore, Margaret. 2017. 'Legitimate Expectations and Land'. *Moral Philosophy and Politics* 4 (2).

Page, Edward A. 2008. 'Distributing the Burdens of Climate Change'. *Environmental Politics* 17 (4): 556–75.

———. 2012. 'Give It up for Climate Change: A Defence of the Beneficiary Pays Principle'. *International Theory* 4 (02): 300–30.

Rawls, John. 1971. *A Theory of Justice*. Original Edition. Cambridge: Belknap Press.

Schüssler, Rudolf. 2011. 'Climate Justice: A Question of Historic Responsibility'? *Journal of Global Ethics* 7 (3): 261–78.

Shepherd, Theodore G. 2014. 'Atmospheric Circulation as a Source of Uncertainty in Climate Change Projections'. *Nature Geoscience* 7 (10): 703–8.

Shue, Henry. 1999. 'Global Environment and International Inequality'. *International Affairs* 75 (3): 531–45.

Stuart-Smith, Rupert F., Friederike E. L. Otto, Aisha I. Saad, Gaia Lisi, Petra Minnerop, Kristian Cedervall Lauta, Kristin van Zwieten, and Thom Wetzer. 2021, June 1–5. 'Filling the Evidentiary Gap in Climate Litigation'. *Nature Climate Change*. https://doi.org/10.1038/s41558-021-01086-7.

Thompson, Janna. 2017. 'Historical Responsibility and Climate Change'. In *Climate Change and Historical Emissions*, edited by Lukas H. Meyer and Pranay Sanklecha, 46–60. Cambridge; New York: Cambridge University Press.

Wallimann-Helmer, Ivo, Lukas Meyer, Kian Mintz-Woo, Thomas Schinko, and Olivia Serdeczny. 2018. 'The Ethical Challenges in the Context of Climate Loss and Damage'. In *Loss and Damage from Climate Change. Concepts, Methods and Policy Options.*, edited by R. Mechler, I. Bouwer, T. Schinko, S. Surminski, and J. Linnerooth-Bayer. Cham, Switzerland: Springer Publishing.

Williams, B. 1973. 'A Critic of Utilitarianism'. In *Utilitarianism: For and Against*, edited by J. J. C. Smart and Bernard Williams. Cambridge: Cambridge University Press.

Wündisch, Joachim. 2017. 'Does Excusable Ignorance Absolve of Liability for Costs'? *Philosophical Studies* 174 (4): 837–51.

———. 2019, August. 'Towards a Non-Ideal Theory of Climate Migration'. *Critical Review of International Social and Political Philosophy* 25 (4): 496–527.

4 Reasons awaiting satisfaction

Calls for climate justice based on a direct principle of historical responsibility are ubiquitous. The words of climate activist Vanessa Nakate for the platform Make Polluters Pay in May 2022 exemplify the spirit of these calls:

> Those who are at the front lines are the least responsible for the climate crisis. I do believe in the Polluter Pays Principle (PPP). Fossil fuel companies and the biggest polluting countries have the responsibility to provide compensation for loss and damage. . . . The Global North hasn't done anything about loss and damage because it has refused to accept responsibility for the climate crisis. They don't want to pay the bill.[1]

Her words resonate with the feeling of many people in the Global South who suffer the consequences of the historical accumulation of emissions, most of which they are not responsible for. The task of this chapter is to provide a justification for the PPP that can preserve its intuitive force as a normative principle for climate justice for loss and damage.

Importantly, this chapter provides a justification for the PPP that can circumvent traditional objections pressed against this principle, based on the knowledge and alternative possibilities conditions of moral responsibility. Respectively, those objections include the Excusable Ignorance Objection, introduced in Chapter 3, and the Path Dependencies Objection, which I will introduce in this chapter. I will leave the explanation for the Causation Objection to Chapter 5, where we delve into how the recent developments in attribution science can contribute to building a rectificatory approach of climate justice for loss and damage.

The justification for the PPP is provided by what I call the Continuity Account. In a nutshell, the Continuity Account claims that polluters should bear the duties of addressing loss and damage because those duties stem from a previously unsatisfied duty of not infringing human rights. In this chapter, I also explain how the account can solve some objections against the PPP and explain how the Continuity Account improves existing accounts to distribute climate change-related duties based on a direct principle of historical responsibility.

DOI: 10.4324/9781003399889-4

4.1 The continuity thesis and the continuity account

The Continuity Account that I develop in this book is based on what John Gardner called the continuity thesis. In a nutshell, this thesis affirms that the grounds for rectificatory justice are the rational remainders of previous unsatisfied moral duties. Admittedly, this thick description requires some unpacking. John Gardner introduced his continuity thesis with the following example:

> I promise to take my children to the beach today, but an emergency intervenes, and I renege on the deal. Let's say I was amply justified in doing so. One of my students, let's say, was in some kind of serious and urgent trouble from which I only could extricate him, and only by devoting most of the day to it. In spite of this ample justification for letting the children down today I am now bound, without having to make a further promise, to take them to the beach at the next suitable opportunity (if there is one). . . Clearly *there is some sense in which my broken promise continues to exert a hold over me* after I break it, *a sense in which it continues to shape what I am bound to do . . . There is an element of continuity here, something that carries through from my original obligation to my obligation now.*
>
> (Gardner 2011, italics are mine).

The point of the continuity thesis is that there is an element of continuity between the initial obligation (e.g., taking my children to the beach) and a newly arising duty (e.g., the duty of taking my children to the beach at the earliest date possible). This element of continuity is constituted by the reasons supporting the actions related to the initial obligation.

The continuity thesis is grounded in the evaluative nature of reasons and the nature of rational agency. Gardner follows Raz in holding that 'reasons for action are . . . evaluative facts, that is, facts consisting in the possession of an evaluative property' (Raz 2003). If someone has a reason to do something, one is compelled to do what one has a reason to do. These 'evaluative properties', Raz claims, are not cancelled out by the contingent situations that may lead to us not conforming to those reasons. Gardner's continuity thesis affirms that the reason one had to perform a certain action (e.g., taking the children to the beach) keeps exerting its pull after having been left unsatisfied. This creates a secondary duty, which is the obligation of compensating the negative effects of not having complied with the initial obligation. Those reasons continue, in a way, obliging the agent and now ground compensatory claim (Gardner 2018)

Let us go back to the promise to take my children to the beach. Because someone promised to take their children to the beach, *their children have a claim on them* to take them to the beach, and *they have an obligation* to take them to the beach. There may have various reasons to comply with this obligation. For instance, one of them is to avoid their disappointment. Since I did not take my children to the beach, they are disappointed. Indeed, one cannot avoid disappointing them anymore. But that does not mean that, having already not complied with that reason, there is nothing one should do. Although one cannot fulfill the promise of taking them to the beach yesterday, one can still do something related to that unfulfilled duty. Not disappointing them was a reason to take them yesterday to the beach and that reason is still awaiting satisfaction today when there is a new opportunity to take them to the beach.

But there are other secondary duties that might emerge from the first one, that is, other ways in which one can conform to the reason not to disappoint them. For instance, imagine that they were not as excited about going to beach as one thought so that they do not want to go to the beach today, nor in the upcoming days or the foreseeable future. However, they might still be disappointed about their parent having broken their promise, and thus one might well have an obligation to apologize and offer to them doing other kind of activity that they might consider fun. Or imagine that they did something even better on that day and now just want to stay at home reading a book. Still, the disappointment they might have suffered when knowing that they would not be taken to the beach as they were promised is a reason for their parent to offer to take them to the beach if they still want to do so. If they do not want, they can 'release' the duty-bearer from that obligation by just declining the offer.

To close this explanation, we should be aware of the stringency of compensatory duties of this sort. Conformity to unsatisfied reasons, both Gardner and Raz admit, does not constitute a *conclusive* reason to satisfy them. Satisfying those reasons might clash with other independent and alternative reasons for action. Those independent and alternative reasons might have more weight than the weight of the previously unsatisfied reasons. In other words, previously unsatisfied reasons only provide an other-things-being-equal duty to conform to them. Admittedly, other reasons might be stronger or more stringent, thereby outweighing the previous ones. However, even in those circumstances, previously unsatisfied reasons and the duty to comply with them do not disappear in thin air. They keep exerting their pull, awaiting for satisfaction at the third, fourth, fifth, etc., best course of action. Thus, the actual rectification of those initial actions depends on the type of moral reasons for actions in place at each time of acting.

The Continuity Account for rectificatory climate justice emerges from the continuity thesis thusly described and justifies the application of a direct principle of historical responsibility, such as the PPP. In order to develop the Continuity Account, we need to come back to the discussion on capabilities and human rights. Recall that I have characterized human rights as specifying 'minimum moral thresholds to which all individuals are entitled, *simply in virtue of their humanity*, and which override all other moral values'. (Caney 2009, 165). As we know well by now, climate change has and will have important negative effects on people's lives. Among other effects, climate change causes severe flooding, intense heatwaves, food insecurity, droughts and increased exposure to vector-borne diseases. Arguably, those events affect many of people's human rights. Following this approach, that people have human rights and that they can be infringed due to climate change constitute moral reasons to *avoid engaging in emission-generating activities* that ultimately cause and contribute to climate change.

The role of the continuity thesis might be then clear. Loss and damage occur when polluters engage in emission-generating activities thereby causing life disruptions that push people below a sufficient level of certain capabilities and the infringement of their human rights. The same moral reasons supporting the avoidance of emission-generating activities keep exerting their hold after they have been left unsatisfied, thereby causing loss and damage. Those unsatisfied moral reasons now ground L&D duties for those who left them unsatisfied by engaging in emission-generating activities, namely, polluters. Thus, polluters have L&D duties because they left unsatisfied the moral reasons speaking against engaging in climate change–inducing activities. Respect for human rights constitute the element of continuity between the first duties (duties to avoid emission-generating activities) and the secondary and newly arising duties (L&D duties). This is what I call the Continuity Account, which supports the PPP for the distribution of L&D duties. In a nutshell, polluting nations should now repair infringements of human rights in the form of loss and damage because they are the result of them leaving unsatisfied reasons against engaging in emission-generating activities, which are the due respect owed to human rights.

As with the continuity thesis, conformity to unsatisfied reasons related to the avoidance of human rights infringements does not constitute a conclusive reason to rectify past behavior. The duty to provide reparations for human rights infringements might clash with other independent and weightier reasons for action. For instance, rectificatory duties might fall on poor or developing countries, with most of their population below sufficiency levels of capabilities, due to their (small) contribution to climate change through emission-generating activities. In some instances, complying with rectificatory duties of this sort might make agents fall below a sufficiency threshold of capabilities, thereby infringing their human rights. However, the Continuity Account would prevent the unwelcome conclusion that, in such circumstances, those agents must comply with their rectificatory duties.

Recall that my account of rectificatory climate justice operates within a sufficientarian background theory of distributive justice, which sets the limits of rectificatory justice demands. Accordingly, claims of rectificatory justice do not outweigh sufficientarian distributive justice considerations. This means that those obliged to rectify climate injustice should not fall below a threshold of sufficiency as a consequence of discharging their rectificatory justice duties. Rectificatory duties that fall on those who find themselves already below a sufficiency threshold or on those who would fall below a sufficiency threshold if those duties where discharged are therefore outweighted by sufficientarian distributive justice considerations. Indeed, avoiding that people fall under the poverty line or not undermining their scare development possibilities are weightier reasons than reasons associated with compensatory duties. For those reasons, my account would most likely select only industrialized high polluting countries as bearers of rectificatory duties. Further, we should not overstate the scale of this objection. Although it is true that poor and developing countries have contributed to climate change, they would also be receiving reparations from an L&D mechanism for the negative effects of climate change. Even if there were assigned reparation duties, in the end, they might not experience any net losses as contributors to such a mechanism.

Finally, one might think that the analogy involving the parent-children relation is not helpful for explaining the duties of a nation and thus cannot support an argument in favor of the PPP based on the continuity thesis. The example of a parent and their children might be too different from the one of polluting nations and victims of loss and damage for the analogy to hold water. One might claim that the first relationship is between two people who share a special relationship, a parent-child relationship, which includes love, care, and responsibility. In contrast, the second relationship is the relationship between a political entity, namely, a nation and individual people. For instance, in this relationship, people do not have the special kind of link existing between parents and children. This difference in relationships, one might argue, comes with a difference in duties.

It is undeniable that parent-children relationships are different from relationships between nations and individual people. However, in the example above, the love and care characterizing the parent-children relationship is not what generates rectificatory duties, even if such a relationship could increase the extend of rectificatory duties or intensify the sense of urgency in complying with them. Rectificatory duties are generated by *the breach of the duty* to abide to one's promises and thus the analogy does not need to rely on the special kind of relationship between the parent and the child. Any example that would involve the breaking of a promise would have similar consequences in terms of resulting in rectificatory duties for those who fail to abide to their promises. Of course, the nature of the rectificatory duties may vary depending on the situation at stake.

One might further object that the disanalogy problems remain, since there is also an important difference concerning the kind of duties involved in each of those cases. The example of the beach above involves duties that stem from a particular voluntary act, namely, the act of promising. When promising to the children taking them to the beach, the parent voluntarily bounds themselves to having a rectificatory duty in case they fail to comply with their promise. In contrast, in the case of climate change, one might argue, there is no voluntary act undertaken by polluting nations that would bound them to the rectification of loss and damage arising from climate change.

Admittedly, the examples differ in terms of how duty-bearers acquire their respective initial duty. Whereas the duty of keeping one promises is acquired through a voluntary action involving making a promise, the duty to respect human rights does not need to involve a voluntary adscription to human rights. The mere existence of human rights is enough for others (including polluting nations) to have the duty to respect individual's human rights and thus avoid their infringements. It is in the very nature of human rights that this kind of voluntary act does not need to be performed for the respective duties to remain applicable. However, there is no reason for this difference to influence the rectificatory duties following the breach of the respective duty. As the continuity thesis points out, the relevant fact for the existence of rectificatory duties is that certain moral reasons attached to our duties have been left unsatisfied when acting in certain ways. Rectificatory duties are there to make up for the failure to satisfy those reasons. Thus, the nature of the initial duties is irrelevant for generating rectificatory duties of the sort discussed in this chapter.

4.2 The continuity account and some objections against the polluter pays principle

In the previous chapter, I introduced two main objections against rectificatory accounts: the Causation Objection and the Excusable Ignorance Objection. I will leave the discussion on the application of the Causation Objection for Chapter 5. Instead, I will focus on the Excusable Ignorance Objection and a third objection that might affect the application of the PPP, namely, the Path Dependencies Objection.

Remember the Excusable Ignorance Objection. The Excusable Ignorance Objection states that no one should be held accountable for the effects of their actions if these were unknown and could not reasonably have been foreseen. Roughly 50% of total emissions have been emitted under conditions of excusable ignorance. Before around 1990, with the publication of the First IPCC Report, there was not enough consensus in the scientific community about the adverse effects of GHG emissions on the climate system. Thus, people and governments could not have reasonably known about the adverse effects of their emission-generating activities. Arguably, non-negligent ignorance undermines attributions of moral responsibility and, thus, complicates justice claims related to the adverse effects of historical emissions.

Note that the Excusable Ignorance Objection relies on the knowledge condition of moral responsibility. That is, the objection relies on the idea that for someone to be morally responsible for something, they need to have had the opportunity to know about the possible negative consequences of their actions. Yet, this is commonly assumed that moral responsibility also requires alternative possibilities: the possibility to act differently than one did (Eshleman 2016). The attribution of moral responsibility to polluters might get complicated also on this front. As we know, early developments of a socioeconomic system condition future pathways. They create technological inertia, institutional designs, and behavioral habits that tend to remain in place for long periods of time because switching to a new system would be too disruptive and expensive (Seto et al. 2016). These mechanisms hinder rapid socioeconomic changes. In that sense, early developments of fossil fuel-based economies created mechanism of carbon lock-in that extended over decades and nudged polluters to continue emitting at very high levels even after knowing about the negative effects of climate change. For that reason, one might argue that high polluting countries cannot be considered morally responsible also for many after-1990 emissions because they did not have a real opportunity to rely on alternative energy sources. Before being considered culpable for the effects of climate change, polluters should be given a 'grace period' to adapt to the new information about climate change and initiate the transition until reaching zero emissions. Until when this grace period can be extended might be a contentious issue. But it is to some extent undeniable that it must be conceded, based on this moral responsibility rationale. Call this objection the Path Dependency Objection.

Thus, The Path Dependencies Objection might constitute another reason to reject the application of the PPP.[2]

The Path Dependencies Objection extends the period for which agents cannot be considered morally responsible for their emissions, thereby challenging the application of the PPP even further. Even if at some point the challenges faced by the PPP will lose strength because they apply to a small set of the overall emissions causing climate change, they are still relevant for an important part of the emissions. Consider that, first, 'excused' emissions based on justified ignorance are still half of the existing overall emissions and, second, changing our socio-economic systems cannot happen in just a couple of decades, for the structural limitations mentioned earlier. For that reason, the application of the PPP for most emissions remains critically challenged (and will be so at least for the near future) by the Excusable Ignorance Objection and the Path Dependencies Objection.

How can the Continuity Account solve these objections? At first glance, it seems that, although moral reasons concerning the protection of people's human rights were left unsatisfied, for long, polluters were not aware of having relevant moral reasons not to engage in emission-generating activities. Hence, the Excusable Ignorance Objection seems to still hold against the PPP. However, against this, note that my argument based on the continuity thesis applies regardless of whether polluters were aware of the negative consequences of their actions, and particularly so in the case of human rights infringements. Recall the first feature attributed to human rights: human rights are grounded *in each person's humanity*. Human beings have certain rights because they have certain relevant properties. As said, one could invoke different set of morally relevant characteristics that human possess and make them worth of having those rights. But regardless of what those exact characteristics are, the idea is that people have these rights because they have these characteristics. Since reasons to respect those rights stem from people having those morally relevant properties, reasons not to infringe their human rights exist as long as people have those characteristics and even in the absence of knowledge concerning the harmful effects of certain actions. Thus, reasons not to infringe human rights were left unsatisfied, even though people could not know about the negative consequences of their emission-generating activities on people's human rights. Those reasons still exert their pull after relevant and reliable knowledge about the effects of climate change becomes available.

Even less contentious should be the circumvention of the Path Dependencies Objection. As I have argued, respect for human rights constitutes a reason not to contribute to the negative effects of climate change and, therefore, a reason to avoid or give up emission-generating activities as much as possible. In circumstances where emitters cannot avoid engaging in emission-generating activities due to carbon lock-in, the moral reasons against emission-generating activities keep exerting their pull. Those reasons await satisfaction even when they are temporarily overridden by reasons to continue emitting or to transition slowly toward low-carbon societies. Ideally, one would satisfy them before the materialization of their possible adverse effects (e.g., by offsetting emissions somewhere else). But when the adverse effects of climate change materialize, satisfying those moral reasons requires repairing their negative effects. Before moving on, let me address one last concern straightforwardly. Some might think that, if the Continuity Account is dependent on polluters not having enough knowledge and alternative paths available to them, we will eventually not need this Continuity Account any more. At the end of the day, there are already some emissions for which polluters are culpable, even if those might not be many. Emissions that were released in the past few years, or in the past few months, might already count as 'culpable emissions', given the existing knowledge about climate change and the alternative socioeconomics paths polluters could have initiated if they really wanted. It is only a matter of time that we can make polluters responsible for their emissions by following a classical understanding of the PPP, which rests on culpable action (Baatz 2013). Even though I would simpatize with a view that rests on culpability for emissions from a certain time onwards, I believe the Continuity Account still has a lot to offer in comparison with a culpability-based approach. For the Continuity Account does not contribute to the blame-game that generates so much rejection among many actors. The Continuity Account is not grounded in culpability and blame, this last one understood as a negative reactive attitude towards those who engage in emissions-generating activities. The Continuity Account is merely meant to attribute responsibility for action based on past involvement in emissions generating activities, but without involving any kind of negative reactive attitudes towards those who participated in them. Even if their action was perfectly justified at that time, they now have a new possibility to act according to the reasons speaking against their past actions. In short, the Continuity Account is focused on responsiveness to reasons rather than on culpability and blame. Thus, the Continuity Account has a lot to offer to justify the *general* application of the PPP if one believes that engaging in a blame-game is not a promising avenue for climate action.

4.3 Alternative accounts

I am not the first author trying to justify the intuitive appeal of a direct principle of historical responsibility. Other philosophers and political theorists before me have provided alternative explanations to the duty of polluting nations to bear the burdens associated with climate change. In this section, I would like to acknowledge their efforts by providing a brief explanation of their accounts. However, I argue that the Continuity Account is better suited to provide the grounds for rectificatory climate justice.

4.3.1 Strict liability

At the beginning of climate ethics debates, various scholars justified the attribution of reparative duties for otherwise blameless emissions based on the concept of strict liability (Shue 1999; Neumayer 2000; Gardiner 2011). Recall from Chapter 3 that a principle of strict liability distributes responsibility for harmful actions regardless of fault. The simple fact that someone causes certain negative outcomes makes this person have the obligation to repair them. Phrases such as 'you broke it, you fix it' or 'clean up your own mess' capture the intuition behind such a principle (Shue 1999; Gardiner 2011). Nevertheless, the fact that many scholars have disagreed with the fairness of such a principle in the context of climate change suggests that this intuition does not find ample support (Schüssler 2011; Moellendorf 2014; Wündisch 2020). Hence, we seem to need more than the appeal to those intuitions.

Gardiner has elaborated further into the justification of a strict liability principle, such as the PPP. This argument relies on certain consequentialist considerations. He believes that a normative system that makes excused ignorants liable for the negative effects of their actions would bring overall benefits. Otherwise, he argues, we would encourage 'phenomena such as "turning a blind eye", self-deception, and cultivated ignorance' (Gardiner and Weisbach 2016, 115). Adopting a general system of strict liability would have good overall consequences because it would incentivize serious research about the potentially bad consequences of our activities.

However, note that this argument does not support *any* application of strict liability. Recall the distinction made in Chapter 3 between prospective and retroactive strict liability. Prospective strict liability is applied when principles of strict liability already exist as part of some regulatory scheme or legal system and their justification lies in the future-orientated implications of these principles. Remember the case of a chemical company polluting a river. Strict liability schemes might be applied in those cases regardless of whether the company made all reasonable efforts to avoid damages and regardless of whether the company had good reasons to believe that these would be the results of its action. The existence and application of such a scheme are justified by the general beneficial consequences of holding companies liable for the harmful effects of their actions when they emit potentially harmful substances. As we saw in Chapter 3, it might well be that making those who cause harm pay for the adverse effects of their action is the way to ensure that they take reasonable steps to avoid harm to happen. Or, in Gardiner's words, it is the best way to avoid that people 'turn a blind eye' or 'cultivate ignorance' regarding the possible negative effects of their actions.

Remember that for the application of strict liability to be justified on these grounds, it is necessary that duty-bearers had the opportunity to be aware of the existence of a scheme of strict liability. The reason is simple: if these principles are to make people maximally careful and discourage reckless conduct, they need to be widely known in advance. Thus, people are aware of the costs they might incur by engaging in certain activities and can decide whether and how to engage in these activities. Otherwise, the disincentivizing mechanism will not work. People can only be motivated by those incentives if they are on notice of the liability attached to the activity (Moellendorf 2014, 168).

However, note that Gardiner wants to use his justification of strict liability (i.e., that the application of strict liability would avoid turning a blind eye) for the application of such a principle, at least partly, to historical emissions, when no strict liability scheme was in place and people were not made aware of the burdens they would be required to bear were their actions to trigger negative effects. That is, Gardiner is using the justification for prospective strict liability to justify a principle of retrospective strict liability, thereby failing to provide an adequate justification. As before, successfully justifying the application of a principle based on its disincentivizing consequences requires that those burdened by the principle know in advance what burdens they would need to bear.

Gardiner justification works for the establishment of a prospective regulatory scheme based on the PPP, whereby future polluters are made to pay for the negative effects of their emissions. However, and importantly, it does not work to justify the application of the PPP neither for historical emissions nor for emissions that would occur until such a regulatory scheme would be in place. Since reparation claims aim at tacking polluting actions that occurred in the past, we still need an argument to support a justification for holding agents retrospectively accountable for the negative effects of their emissions. My Continuity Account can do this work. This account explains that someone has a duty to satisfy previously unsatisfied moral reasons for acting in certain ways. In the case of loss and damage, those moral reasons referred to the duty to avoid contributions to climate change in the form of emission-generating activities because of their emissions on human rights infringements. When agents leave those reasons unsatisfied, they have the duty to satisfy them at the next possible occasion, in this case by providing reparations for loss and damage. My account does not preclude that a system of strict liability is also implemented for forward-looking reasons, having to do with preventing the kind of attitudes Gardiner's account aims to prevent, namely, that agents turn a blind eye on certain risks their action may cause. However, the account provides the groundwork for justifying retrospective strict liability principle, such as the one required to justify the application of the PPP.

4.3.2 *Counterfactual liability*

An alternative account, prominently proposed by Daniel Butt, relies on the notion of counterfactual liability (Butt 2017). Counterfactual liability applies to actions for which agents would not be held liable if it weren't for the belief that the absence of exculpatory circumstances would not have changed their behavior. Thus, for counterfactual liability various conditions need to be met. First, liable agents must have caused certain negative effects. Second, the situation involves certain factors that, under normal circumstances, would exempt them from liability. But, third, there are good reasons to believe that, in the absence of those exculpatory factors, those agents would have undertaken those actions anyway. Under those conditions, agents can be made 'counterfactually liable' for the negative effects of their actions.

When applied to historical emissions, the full argument based on counterfactual liability affirms that polluters are counterfactual liable for the negative effects of climate change (first condition) even in the presence of excusably ignorant factors and path dependencies (second condition) because, even if they had known about the negative effects of their emissions, and even if they could have transited faster to low-carbon societies, they would have not done so (third condition). Therefore, polluters should be held counterfactual liable for their emission-generating activities.

Butt presents his counterfactual liability argument as a response to the excusable ignorance objection. His argument has an empirical dimension and a normative dimension. His empirical claim is that had historical emitters known about the negative consequences of GHG emissions before 1990, they would have relied on emission-generating activities in the way they did anyway.

This claim is based on two facts: the general behavior of industrialized countries toward other nations and their post-1990 emitting patterns. First, according to Butt, the history of the relations between industrial powers and people living in developing countries is full of disregard and mistreatment, including the period when GHGs emissions started to grow exponentially. 'The dominant mode of foreign policy for much of the period in question was imperial and/or colonial: many less developed countries were subject to grievous wrongdoing at the hands of Western countries eager to fuel their industrial growth' (Butt 2017, 67). His argument is that if industrial powers were willing to subjugate people out of sheer economic self-interest, they would have been willing to engage in climate change–inducing activities from which they could make economic profit, regardless of whether that had negative impacts on others. That is, knowledge about the negative effects of climate change would not have prevented them from engaging in high polluting activities. We can summarize this argument by invoking the distinction between acting *while* ignorant and acting *from* ignorance (Aristotle 2004; Zimmerman 1997). Although historical emitters might have acted *while* ignoring the negative effects of their emission-generating activities, they did not act *from* that ignorance. That is, their ignorance is not the explanatory fact for why they acted in the way they did.

Second, this argument is reinforced by evidence on how industrial powers acted after knowing about the negative effects of their emissions. As Butt argues, industrial powers did not change their polluting behavior over a long period of time after the publication of the First IPCC report in 1990. Although one could argue that this might be because of the previously mentioned path dependencies and carbon lock-in mechanisms, it is also true that their lack for climate action had lasted long after they got to know about the negative effects of climate change and their timid changes cannot be explained *only* by that inertia. One could argue that at some point, at least, industrial powers should have acted more promptly than they did, which shows their lack of willingness to take seriously the suffering of those negatively affected by climate change. From these empirical considerations, Butt draws the conclusion that:

> [W]hen we look . . . at the (general) moral character of historic communities along with the evidence of how these communities reacted when they became aware of the effects of their actions . . . it is, to the very least, very hard to make a good faith argument that things would have been different had scientific knowledge been more advanced at an earlier date.
>
> (Butt 2017, 69)

Butt's argument based on counterfactual liability affirms that polluters are counterfactual liable for the negative effects of climate change because, even if they had known about the negative effects of their emissions, they would have not done so. This counterfactual account does rely on attributions of moral responsibility, blameworthiness, and culpability. Instead, that negative appraisal only grounds a normative argument assigning remedial responsibility. Here, 'remedial responsibility' refers to the duty of repairing the negative effects of climate change. 'The point of considering the moral character of the action is to determine not moral but *remedial responsibility*' (Butt 2017, 70).

Note that Butt's counterfactual liability account could be likewise applied to cases where the exculpatory circumstances are not based on excusable ignorance but on path dependencies. Similarly, two empirical facts could support the attribution of counterfactual liability: the general behavior of industrialized countries toward other nations and their general emitting patterns. The behavior of industrialized nations toward other nations shows the general disrespected that has characterized their historical conduct. Further, even if those countries were trapped in some path dependencies and lock-in mechanisms, the world witnessed an increase in industrialized countries overall emissions for many years after some actions started to be taken. These increases can be hardly explained only by initial inertia (Shue 1999; Gardiner 2011). Even though that tendency has been somewhat reversed, the development and technological capacity of industrialized countries would provide a reason enough to believe that they could have done much more to revert or slow down the growing tendency of emissions rates at an earlier stage. Hence, one might conclude, following Butt's counterfactual reasoning, that even if polluters could have transited earlier to low-carbon societies (thereby not being affected by path dependencies), they would have not done so at the required speed.

Summarizing, the full argument based on counterfactual liability affirms that polluters are counterfactual liable for the negative effects of climate change because, even if they had known about the negative effects of their emissions, and even if they could have transited faster to low-carbon societies, they would have not done so. This counterfactual account does rely on attributions of moral responsibility, blameworthiness, and culpability.

The account based on counterfactual liability account is not free from objections. First, Butt's quasi-empirical remarks are contested. The behavior of historical emissions in certain regards (i.e., concerning colonial actions) does not show straightforwardly how they would have acted in other regards (i.e., concerning emission-generating activities) had they had the relevant knowledge about climate change. Arguably, there are relevant differences between colonial actions and climate change–inducing activities that cast some doubt on this claim. For instance, whereas colonial actions only negatively affected people living in the colonies (and not even everyone living in the colonies), the negative effects of excessive emissions can also affect those engaging in those activities (i.e., high emitters). Although one might argue that emissions levels could be limited not to affect highly industrialized countries in the Global North, the negative effects of climate change are not that easy to foresee. Since they can be potentially very dangerous, it is doubtful whether industrial powers would have nonetheless engaged in those activities had the known about their potentially dangerous effects. Furthermore, their post-1990 behavior could have different explanations other than their predisposition to disregard others. Industrial powers might have continued emitting, even at higher levels, because the path dependencies and lock-in mechanism developed due to their long-term reliance on fossil fuel–related infrastructures and related investments. Therefore, it is difficult to draw firm conclusions about how historical emitters would have acted.

More importantly, there are some procedural justice worries with counterfactual liability. One might argue that these kinds of arguments contravene procedural justice because they made judgments based on what people would have done, rather than for what they actually did (Caney 2010). Arguably, this is procedurally unjust since people should be judged for what they do and not for what they would have done had the circumstances been different, and especially so given the aforementioned difficulties in drawing firm conclusions from the empirical facts presented earlier.

In contrast, the Continuity Account does not fall into these problems. First, this account does not rely on contested hypothesis about how the world would have been if industrialized countries had had more knowledge about their climate change inducing actions. Also, it does not judge them for what they would have done in a different counterfactual situation. Even if it would be true that historical emitters would have anyway acted in the ways they did had they known all the relevant facts, their emission-generating activities contributed to cause human rights infringements. That fact alone constituted a moral reason not to act in the way they did and that provides a secondary reason to act in ways that satisfy those reasons, now by repairing the negative effects of climate change. In other words, their contribution to human rights infringements alone grounds a duty to rectify the effects of not having satisfied reasons against contributing to human rights infringements in the first place. Moreover, this conclusion should not trigger procedural justice concerns since the duty-bearers are not judged for things they did not do, but only for things they did, namely, contributing to human rights infringements.

4.3.3 Outcome responsibility

The latest account relies on the concept of outcome responsibility. Outcome responsibility is a non-fault-based notion of responsibility associated with actions undertaken under conditions of known unknown risks, that is, when we act knowing that there are certain risks that we are not fully knowledgeable of. Alexa Zellentin has relied on the notion of outcome responsibility to the defend the application of the PPP to responsibilities associated with the negative effects of climate change.[3]

Zellentin argues that, when we act, we either know that our actions fall under our secure sphere of competence, or we do not know that our actions fall under our secure sphere of competence. For the first kind of actions, we know their likely results, as well as that we can undertake them carefully enough without causing negative outcomes. For the second kind of actions, we don't know the potential results. That is, we know that we do not have the relevant competence to know all the relevant facts concerning the possible negative results and we are aware that, precisely because of our incompetence, negative outcomes could occur. In such a situation, Zellentin argues, if we act, we make an 'implicit bet': while being aware that we might be undertaking risky actions, we *bet* that our action would not result in negative outcomes. Making that implicit bet is what makes us outcome responsible if the potentially negative effects of our actions were to materialize. But this also has a normative dimension since in those circumstances, one should accept bearing the eventual burdens associated with possible bad outcomes. In Zellentin's words: 'my taking the bet involves acknowledging the decision to act and thus acknowledging that there is a normatively significant link between me and the consequences of my acts that might generate rectificatory duties toward those whose rights I accidentally infringe' (Zellentin 2015, 264).

Zellentin applies this notion to the context of historical emissions. She argues that historical emitters are outcome-responsible for pre-1990 emissions because they were (or should have been) aware that the potential risks of their polluting activities were not fully understood. At the end of the 19th century, scientists had already pointed out that greenhouse gases may not be overall beneficial and that industrialization had consequences on a level and to an extent that was yet to be explored (Arrhenius 1896). Moreover, around that time, novelists such as Charles Dickens reflected their concerns about the impact of large-scale industrialization on the environment (Dickens 1854/2004). Thus, they knew or at least should have known that their emissions could have negative consequences that they did not fully understand. Since they decided to carry on with their activities, they should now bear the burdens of repairing their negative effects.

However, if outcome responsibility relies on making this kind of implicit bet, there is no reason to restrict its application to actions that fall outside of our secure sphere of competence. Note that when we act *within* our secure sphere of competence, we undertake actions that we are usually competent to undertake with reasonable and sufficient precautions. Here, we also make an implicit bet: We bet that our capacities are as good and enough not to cause any accident. But accidents may always happen. The only difference is that the likelihood of a negative outcome is lower. In fact, Zellentin implicitly recognizes this point in her last contribution to the topic (Zellentin 2018). She argues that her notion of outcome responsibility also applies to cases where 'we are in the midst of doing something that we know we are usually well able to accomplish, but our capacities fail' (Zellentin 2018, 11). She argues that, in those cases, outcome responsibility also applies because the practice of attributing responsibility should rest on 'an understanding of *agency* that is informed by the knowledge that human beings are fallible beings' (Zellentin 2018, 12). This second case also involves an 'implicit bet': when we undertake actions that fall under our secure sphere of competence, we also bet that we will not cause any negative outcome.

Arguably, the notion of outcome responsibility based on making an implicit bet could also be applied to answer the Path Dependencies Objection. The reasoning could be roughly as follows. When one chooses a pathway (i.e., a socioeconomic pathway based on the use of fossil fuels), one makes an implicit bet: one bets that this pathway will be overall successful, in terms of both bringing positive outcomes and avoiding major negative outcomes. Following Zellentin's line of reasoning, in those circumstances, one should accept bearing the eventual burdens associated with possible bad outcomes. In this case, accepting such burdens involves repairing the negative effects produced by socioeconomic pathways based on the use of fossil fuels. Thus, the notion of implicit bet could also be helpful to explain why polluters have rectificatory duties toward the victims of loss and damage even if they were affected by path dependencies and lock-in mechanisms. Even if one could not avoid causing those harms, initiating that pathway involved an implicit bet that now creates certain normative obligations.

Initially, the formulation 'acting outside of the secure sphere of competence' seemed to rely on agents making choices *knowing* that their knowledge was insufficient for acting 'safely'. Thus, this formulation grounds rectificatory duties on the undertaking of excessive risks due to the awareness about having insufficient information. But Zellentin is careful enough not to take this route. If polluters were aware or could have been aware about having insufficient information to act safely, then they were not excusably ignorant: they knew or could have known that they were not competent to act under the required levels of safety. In this situation, instead of engaging in emission-generating activities, they should have researched further about the negative effects of their polluting activities. That is, if this were her argumentative route, Zellentin's argument would not show that the Excusable Ignorance Objection *applies* (i.e., polluters were excused for their ignorance) but *nonetheless* should be held accountable for the negative effects of their emissions. Instead, her argument would show that the Excusable Ignorance Objection does *not apply* in this context, that is, that polluters should not be excused for their ignorance because they knew enough about their ignorance not to engage in emission-generating activities without further research.

However, Zellentin's empirical evidence is too thin for this argumentative strategy to be successful. Arguably, Arrhenius' experiments suggesting that GHGs might not been overall beneficial and the observation of a novelist about polluting activities being bad for the local environment are not enough to claim that polluters should have engaged in more research before continuing with their emission-generating activities. At least, most philosophers would disagree with such a claim, given that the excusable ignorance period is usually taken to be the publication of the first IPCC report in 1990. The difference between those two epistemic landmarks is too big to think that unexcused ignorance could start even remotely close to what Zellentin's argument, under this interpretation, would suggest.

For those reasons, I suspect, Zellentin's account does not rely on polluter's epistemic state but simply on the fact that they engaged in polluting activities. It is the *acting*, not the epistemic conditions under which one acts, what confers rectificatory duties. Recall that she states that

> my taking the bet involves acknowledging the decision to act and thus acknowledging that there is a normatively significant link between me and

> the consequences of my acts that might generate rectificatory duties toward those whose rights I accidentally infringe.
>
> (Zellentin 2015, 264)

However, as the attentive reader might have already realized, there is an unexplained gap between the decision to act and the generation of rectificatory duties on this account. Why should simply acting generate rectificatory duties?

I do not deny that such an explanation could be provided under an outcome responsibility account. However, identifying this unexplained gap enables us to see the main advantage of my Continuity Account over this alternative account based on outcome responsibility. Unlike the outcome responsibility account, the Continuity Account *does* provide an explanation for why acting in ways that infringe human rights generates rectificatory duties. As we saw, reasons for climate reparations are grounded in the existing duty to satisfy moral reasons based on due respect for human rights that are left unsatisfied when one engages in emission-generating activities.

4.4 Conclusion

Calls for climate justice made by climate activists, poor countries, and civil society in industrialized countries alike have appeal to the direct historical responsibility of polluters to make up for the negative effects of their emissions. However, principles of historical responsibility, such as the PPP, have been dismissed by a wide range of political representatives of industrialized countries, historical polluting industries, philosophers, and political theorists. The task of this chapter has been to provide a justification for the PPP as a direct principle of historical responsibility that can preserve its intuitive force as a normative principle for climate justice for loss and damage.

The justification I provided for the PPP in this chapter is based on the continuity thesis, which I named the Continuity Account. This account claims that polluters should bear the burdens of addressing loss and damage as a reaction to the previously unsatisfied moral reasons not to infringe human rights. Due respect for human rights grounds the general duty to avoid actions leading to human rights infringements. Climate change–inducing activities fall below the category of actions that ought to be avoided for those reasons. The Continuity Account states, based on the continuity thesis, that polluting agents have the duty to rectify the negative effects of climate change based on due respect human rights, which constitutes a moral reason to avoid engaging in emission-generating activities in the first place. When those moral reasons are left unsatisfied, polluters have the obligation to rectify their contribution to human rights infringements by providing reparations for loss and damage of the sort discussed in Chapter 3.

In this chapter, I have also explained how the Continuity Account can circumvent two important objections affecting a direct principle of historical responsibility, such as the PPP: the Excusable Ignorance Objection and the Path Dependencies Objection. In a nutshell, I have argued that due respect for human rights is a duty that depends neither on the epistemic status nor on the alternative actions available to those who failed to abide to it. For those reasons, the Continuity Account can circumvent those two objections. Even more, the Continuity Account can provide a 'blame-free' narrative for assigning historical responsibility for climate change.

This chapter also explained how the Continuity Account improves the existing accounts supporting a direct principle of historical responsibility. First, unlike the prospective justification for a principle of strict liability, the Continuity Account can justify the imposition of burdens to address loss and damage for emissions that occur before the implementation of a general liability scheme. Second, unlike the account based on counterfactual liability, the Continuity Account does not fall into procedural justice problems concerning the distribution of burdens based on questionable assumptions about how duty-bearers had acted in alternative hypothetical circumstances. Finally, the Continuity Account fills the explanatory cap left by an account based on outcome responsibility by elaborating on why polluters ought to rectify climate change–inducing actions that lead to human rights infringements.

Notes

1 The acronym PPP in the quotation was added. 'Make Polluters Pay' is a coalition of charitable and campaigning groups and organizations, campaigning for loss and damage finance, paid for by polluters. For more information, see: https://makepolluterspay.co.uk/make-polluters-pay. Similar calls have been made by organizations campaigning for climate justice, such as Oxfam. For more information, see: https://www.oxfam.org.uk/get-involved/campaign-with-oxfam/tax-the-biggest-polluters-now/#2e30cf7c-b26b-4c5d-8906-7c4d069d9718-who-are-the-polluters-that-should-pay.

2 Note that the BPP is not affected by this objection since beneficiaries, by assumption, are not held responsible for their actions. For this reason, I left the discussion of the Path Dependencies Objection for this chapter.

3 Originally, the concept of outcome responsibility was presented by Tony Honoré (1999) and developed further by David Miller (Miller 2007). Here, I only discuss the notion of outcomes responsibility as presented by Zellentin. I do not address whether this use of the concept of outcome responsibility correctly represents that of Honoré and Miller.

References

Aristotle. 2004. *Nicomachean Ethics*, edited by Roger Crisp. Cambridge: Cambridge University Press.

Arrhenius, Svante. 1896. 'XXXI. On the Influence of Carbonic Acid in the Air upon the Temperature of the Ground'. *The London, Edinburgh, and Dublin Philosophical Magazine and Journal of Science* 41 (251): 237–76. https://doi.org/10.1080/14786449608620846.

Baatz, Christian. 2013. 'Responsibility for the Past? Some Thoughts on Compensating Those Vulnerable to Climate Change in Developing Countries'. *Ethics, Policy & Environment* 16 (1): 94–110.

Butt, Daniel. 2017. 'Historical Emissions: Does Ignorance Matter'? In *Climate Justice and Historical Emissions*, edited by Lukas H. Meyer and Pranay Sanklecha, 61–79.1st Edition. Cambridge University Press. https://doi.org/10.1017/9781107706835.004.

Caney, Simon. 2009. 'Climate Change, Human Rights and Moral Thresholds'. In *Human Rights and Climate Change*, edited by Stephen Humphreys. Cambridge; New York: Cambridge University Press.

———. 2010. 'Climate Change and the Duties of the Advantaged'. *Critical Review of International Social and Political Philosophy* 13 (1): 203–28.
Dickens, Charles. 2004. *Hard Times*. Harmondsworth: Penguin Classics. (Original work published 1854).
Eshleman, Andrew. 2016. 'Moral Responsibility'. In *The* Stanford *Encyclopedia of Philosophy*, edited by Edward N. Zalta, Winter. Stanford: Metaphysics Research Lab, Stanford University. https://plato.stanford.edu/archives/win2016/entries/moral-responsibility/.
Gardiner, Stephen. 2011. *A Perfect Moral Storm: The Ethical Tragedy of Climate Change*. Environmental Ethics and Science Policy Series. New York: Oxford University Press.
Gardiner, Stephen M., and David A. Weisbach. 2016. *Debating Climate Ethics*. *Debating Ethics*. New York: Oxford University Press.
Gardner, John. 2011. 'What Is Tort Law for? Part I. The Place of Corrective Justice'. *Law and Philosophy* 30 (1): 1–50.
———. 2018. *From Personal Life to Private Law*. Oxford: Oxford University Press.
Honoré, Tony. 1999. *Responsibility and Fault*. Oxford; Portland: Hart Publishing.
Miller, David. 2007. *National Responsibility and Global Justice*. Oxford; New York: Oxford University Press.
Moellendorf, Darrel. 2014. *The Moral Challenge of Dangerous Climate Change: Values, Poverty, and Policy*. New York: Cambridge University Press.
Neumayer, Eric. 2000. 'In Defence of Historical Accountability for Greenhouse Gas Emissions'. *Ecological Economics* 33 (2): 185–92.
Raz, Joseph. 2003. 'Personal Practical Conflicts'. In *Practical Conflicts: New Philosophical Essays*, edited by Peter Baumann and Monika Betzler. Cambridge: Cambridge University Press.
Schüssler, Rudolf. 2011. 'Climate Justice: A Question of Historic Responsibility'? *Journal of Global Ethics* 7 (3): 261–78.
Seto, Karen C., Steven J. Davis, Ronald B. Mitchell, Eleanor C. Stokes, Gregory Unruh, and Diana Ürge-Vorsatz. 2016. 'Carbon Lock-In: Types, Causes, and Policy Implications'. *Annual Review of Environment and Resources* 41 (1): 425–52.
Shue, Henry. 1999. 'Global Environment and International Inequality'. *International Affairs* 75 (3): 531–45.
Wündisch, Joachim. 2020. 'Middle Ground on Liability for Costs'? *Philosophical Studies* 177 (10): 3097–115.
Zellentin, Alexa. 2015. 'Compensation for Historical Emissions and Excusable Ignorance'. *Journal of Applied Philosophy* 32 (3): 258–74.
———. 2018. *Outcome Responsibility: Fallible Beings Acting in an Uncertain World, SPIRe Working Paper Wp07*. Dublin: Schold of Politics and International Relations, UCD.
Zimmerman, M. J. 1997. 'Moral Responsibility and Ignorance'. *Ethics* 107 (3): 410–26.

5 Climate harm and attribution science

Chapter 3 elaborated on the definition of loss and damage originally provided by Page and Heyward. Remember that, for them, loss and damage ought to be considered 'the unjustified disruptions in the lives of individuals and communities, whether permanent or otherwise, *that are attributable to anthropogenic climate change* and which remain after mitigation and adaptation efforts have been attempted' (Page and Heyward 2016, 3, italics mine). Back then, I delved into how we should understand those life disruptions and the measures to address them. However, I left a missing piece concerning how those life disruptions ought to be attributed to climate change for them to count as loss and damage. That is, that chapter left unanswered how to identify climate harm. Likewise, I left this problem unaddressed in Chapter 4, when leaving aside the Causation Objection. This chapter gets back to this issue and addresses this problem.

Isolating the domain of climate justice for loss and damage requires identifying a connection between environmental loss and damage and emission-generating activities. Political philosophers working on climate justice have avoided answering the questions about how this identification would be possible, sometimes because they wrongly thought that their principles of climate justice would not require establishing such a connection, or sometimes because they thought that other more loose connections would be enough. In Chapter 3, I have already replied to the first ones by showing why answering the Causation Objection is relevant to preserve the independency of the climate justice domain: in a nutshell, because it enables us to differentiate between victims of any other injustice and victims of climate change.

This is not to say that we can achieve a perfect identification of the victims of climate change. Given the complexity of climate change, we can probably only aspire to provide the best available empirical information to establish a connection between environmental loss and damage and climate change. Moral and political philosophers have not done this work so far probably given the complexities involved in understanding climate science and its potential usefulness for political purposes. The space of this book is probably not enough for me to develop this task to the fullest. However, we need to start somewhere. The next two chapters aim to advance this necessary work. In this chapter, I present the two most prominent methodologies for attributing weather events to climate change and assess whether

DOI: 10.4324/9781003399889-5

we have reasons to reject one of them for being comparatively more unfair toward potential duty bearers, as some critics have implied. I conclude that we do not have reasons to reject any of these attribution methods in favor of the other based on those reasons. This question is relevant because, if it were not so, we would have a reason to prefer one attribution method over the other, when it comes to using these attribution methods for climate justice purposes. But I conclude that we do not have reasons to do so. Building on this result, in Chapter 6, I assess which of those attribution methods are more purpose for a policy mechanism for loss and damage. Although my discussion will not fully settle the matter, it constitutes an important step forward to investigating how to use climate science for climate justice for loss and damage.

We know that anthropogenic greenhouse gas emissions and other human activities are a major forcing of recent climatic changes. But we are less certain about the link between particular EWEs and anthropogenic forcing since EWEs would occur even in a preindustrial climate (IPCC 2021, AR6, chapter 11, sections 2.3 and 2.4). However, this link seems to be of particular interest because changes in local weather affect societies more directly (Allen 2012; Stott et al. 2016; Nature Editorial 2018). Particularly, the link between ACC and specific EWE is relevant for advancing justice claims related to loss and damage (Thompson and Otto 2015).[1]

To solve the Causation Objection, and thus link environmental loss and damage with climate change, climate scientists propose different attribution methods to establish a link between EWEs and ACC. The first methodology to emerge was the probabilistic approach – also known as probabilistic approach, probabilistic event attribution or just PEA (Allen 2003; Stott, Stone, and Allen 2004; Stott et al. 2013; Stott et al. 2016; Otto et al. 2017).[2] Later, another group of scientists offered an alternative based on conditional attribution (Trenberth, Fasullo, and Shepherd 2015), which is a specific application of a specific type of storyline approach (Shepherd 2016; Shepherd et al. 2018). In line with other publications in the philosophy of science literature, I refer to this approach as the storyline approach.

The choice between the probabilistic and the storyline approach must take into consideration the features of each of those approaches. Particularly, this choice should assess how their features might affect fairness considerations in the attribution of responsibilities for climate change loss and damage. In this context, it is relevant to analyze the main worry that probabilistic approach defendants have raised against the storyline approach. This worry has been that the storyline approach could be overstating the role of ACC in EWEs (Allen 2011; Stott et al. 2013; Stott et al. 2016; Stott, Karoly, and Zwiers 2017). In fact, this worry has led to a general association of the storyline approach with overstatements of ACC, in contrast to the probabilistic approach.

Whether or not an approach overstates the effects of ACC might be relevant in different climate policy contexts, but above all in loss and damage and adaptation contexts. Those stakeholders who seek to avoid overstating the effects of ACC might base their decision-making on the results provided by PEA studies instead of studies carried out with the storyline approach. For instance, this categorization might be relevant in liability contexts for loss and damage. If stakeholders seek to avoid being partial for or against one of the parties, they might forgo using an attribution method that presumably overstates the effects of ACC because such a method might be partial against the putative liable parties. These might constitute reasons to prefer an alternative approach, that is, the probabilistic approach, that presumably would not overstate the effects of ACC.[3]

Hence, the association between the use of the storyline approach and overstatements of ACC and how this association affects the choice between different attribution methods deserve further clarification and investigation if attribution science is going to be able to establish a connection between EWEs and ACC that solves the Causation Objection.

In this chapter, I explore a variety of research questions related to this issue. First, whether the storyline approach necessarily overstates the effects of anthropogenic climate change. Second, whether the objections offered against the storyline approach constitute good reasons to prefer the probabilistic approach. In a nutshell, I argue that the storyline approach does not necessarily overstate the effects of ACC and that the objections offered against this approach cannot constitute good reasons to prefer, in general, the probabilistic approach because this one is often affected by the same or very similar objections.

5.1 Attribution methods: the probabilistic and the storyline approach

Generally, attribution science aims at identifying in which sense and/or to what extent a certain EWE can be attributed to ACC. However, this general question can be interpreted in at least two different ways. The probabilistic approach and the storyline approach differ precisely in the way they approach that general question.

The probabilistic approach has so far been the conventional methodology in attributing EWEs to human forcing (Allen 2003, 2011; Stott. et al. 2013; Stott et al. 2016; Otto et al. 2017). This methodology takes a certain event as a token of a class of EWEs and asks the following research question: How much did ACC increase the probability or risk of a specific type of event? Answering this question requires comparing the probability (p1) or risk of a specific class of events in a world affected by ACC (actual world) and the probability (p0) of such a type of event in a world without ACC (counterfactual world).

The result of this process would be statements of this sort: 'ACC has increased the probability of occurrence of this type of EWE by a factor X (probability ratio)'. Or, put in other words: 'an event of that class was X% more likely to occur in a world with ACC than in a world without ACC'. This operation is often represented as a Fraction of Attributable Risk (FAR), where FAR = 1 – (p0/p1). The FAR is interpreted as the fraction of the risk of an event that is attributable to the external forcing. Hence, FAR leads to probabilistic causal claims, such as: 'it is very likely that X amount of the risk of this EWE is attributable to anthropogenic forcing'. In sum, the probabilistic approach attributes a fraction of the probability or the risk of the event occurring to ACC.

Of course, it is also possible that this methodology shows a decrease in the probability of the EWE occurring due to ACC. Therefore, statements of the sort 'ACC has decreased the probability of occurrence of this type of EWE by a factor X' or 'an event of that class was X% less likely to occur in a world with ACC than in a world without ACC' are also possible. Although scientists working within this approach are less prone to emphasize this point, it is important to bear in mind that the influence of ACC on the EWE can also work in the opposite direction and decrease the probability of the event occurring. It is also possible that the probability of the event is unaffected by ACC and remains essentially the same.

Importantly, the PEA community has suggested that the results of their studies could support the attribution of the harmful effects of EWEs to ACC when certain thresholds are reached, as is done in tort law contexts (Stott, Stone, and Allen 2004; Allen et al. 2007; Allen 2012; Thompson and Otto 2015; Otto et al. 2016). In tort law contexts, factor X can be said to have caused an effect Y if, on the balance of probabilities, it is more likely than not that X caused Y (Lloyd et al. 2021). For this to be the case, X must have more than doubled the probability of Y occurring. Famously, this threshold has been successfully applied in asbestos lawsuits, where certain negative health conditions were attributed to the use of asbestos because it was considered that this substance more than doubled the probability of their occurrence. With this in mind, many scholars have suggested that a 0.5 FAR could be considered a relevant threshold for attributing the harmful effects of EWEs to ACC, which can be relevant in liability contexts (Grossman 2003; Pall et al. 2011; Allen 2012; Hannart et al. 2016; Stuart-Smith et al. 2021).[4]

The application of the storyline approach to attribution studies emerges from certain skepticism concerning the application of probability assessments to a certain type of EWE.[5] To explain this skepticism, proponents of the storyline approach point out the different contributions of dynamic and thermodynamic climate variables to an EWE (Trenberth et al. 2015; Shepherd et al. 2018). Simplifying, dynamic factors include specific weather patterns such as cyclonic storms or persistent blocking highs that are responsible for the occurrence of a given weather event at a given time. Thermodynamic factors include, for example, surface warming and moistening of the atmosphere and strongly influence the *severity* of an EWE.

For the thermodynamic aspect of the event, models typically simulate robust changes in a warming climate, but changes in atmospheric dynamics are usually much more uncertain. Indeed, Shepherd affirms that:

> the most uncertain aspect of climate modeling lies in the representation of unresolved (sub-gridscale) processes such as clouds, convection, and boundary-layer and gravity-wave drag, and its sensitive interaction with large-scale dynamics. It is, therefore, reasonable to hypothesize that the representation of these processes is responsible for systematic non-robustness of the predicted circulation response to climate change.
>
> (Shepherd 2014, 706)

For these reasons, defenders of the storyline approach argue that identifying a possible human contribution to changes in dynamic climate variables is very challenging and it often delivers unreliable or inconclusive results. Similarly, Trenberth et al. argue that 'although large changes in atmospheric circulation can be readily apparent in a single climate model run, they are not robust and can change considerably in the next run or model', and, importantly, that 'forced circulation changes are not well established, and it is difficult to detect changes in circulation-related extremes in observations because of small signal-to-noise ratios' (2015, 725).

The problem is that, because the PEA community wants to consider the EWE as a 'single, self-reinforcing and indivisible whole' (Allen 2012, 13), the probabilistic approach aims to track both dynamic and thermodynamic changes in the EWE due to ACC. But this would be problematic, according to proponents of this approach, because of the cited challenges in representing dynamic changes. Hence, these scientists conclude that

> the conventional approach to extreme event attribution [PEA] is rather inefficient in cases that are strongly governed by changed circulation, with a generally inconclusive outcome. Even when a detectable anthropogenic influence is found in a model, the reliability of that finding cannot carry much weight.
> (Trenberth et al. 2015, 726)

In a nutshell, the criticism is that these problems might make attribution studies miss the effects of ACC and also undermine their reliability.[6]

However, scientists working on the storyline approach do not refuse to attribute EWE (or, at least, some aspects of these events) to anthropogenic forcings.[7] Instead of focusing on types of EWE, the storyline approach focuses on concrete events and investigates their sources in a conditional manner (Shepherd 2014, 2016). Shepherd has described the storyline approach as 'analogous to accident investigation (where multiple contributing factors are generally involved and their roles are assessed in a conditional manner)' (Shepherd 2016, 32). To do so, scientists proceed by taking the large-scale dynamic state of an event as a given constraint and then ask about the contribution of human forcing to the event's thermodynamic climate variables. In that way, they obtain answers to the attribution question conditioned on the given dynamic components (see, for instance, Pall et al. 2017; Patricola and Wehner 2018; Takayabu et al. 2015; Sillmann et al. 2021; IPCC AR6 Ch 11, Section 11.2.3). This approach is less prone to errors related to the unreliability of climate models because it does not depend on the ability of these models to simulate changes in atmospheric circulation.

After fixing the dynamic variables, the storyline approach shifts the research question into one about the event's magnitude or severity. Instead of asking how much anthropogenic forcing has increased the likelihood of the event happening, this approach focuses on the effects of anthropogenic forcing in the increase of the event's magnitude. That is, the relevant research question is not: 'how much has anthropogenic forcing increased the likelihood of the event happening?' Instead, the question is: 'Given certain dynamic variables, how much has ACC increased the magnitude of this particular event?' Accordingly, the answers attached to this methodology are of this sort: 'ACC increased the EWE's magnitude by a value or factor of X'. As before, the methodology can also show a decrease in the magnitude, thereby leading to statements such as: 'ACC has decreased the EWE's magnitude by a value or factor of X'.

5.2 The reaction and criticism of the PEA community toward the storyline approach

The emergence of the storyline approach triggered some controversy within the scientific community, and especially among the PEA community. Their main complaint is that the conditional structure of the storyline approach and its focus on thermodynamics could make this approach overstate the effects of ACC.

This worry has been captured in various papers from different proponents of the PEA community. For instance, Stott et al. argued that 'by always finding a role for human-induced effects, attribution assessments that only consider thermodynamics could overstate the role of ACC' (2016, 33; similarly in Stott, Karoly, and Zwiers 2017, 147). Similarly, Allen (2011) accused some proponents of the storyline approach – in particular Trenberth (2011) and Curry (2011) – of assuming that ACC had always an impact on local weather events and that this puts scientists at risk of making false-positive errors. The spirit of this complaint was largely in line with that of Stott and colleagues, namely, that the new alternative approach could be overstating the effects of ACC.[8] These complaints also imply that the probabilistic approach might be less prone to overstate the effects of climate change and, if anything, it has the opposite tendency, thereby position it more in line with values of scientific rigor.[9]

Some philosophers of science have interpreted this criticism as suggesting a tendency of the storyline approach toward overstatements. Accordingly, the criticism of the PEA community would be that the storyline approach is *prone* to overstating the effects of ACC (Lloyd and Oreskes 2018, 2019; Winsberg, Oreskes, and Lloyd 2020; Pulkkinen et al. 2022). However, the PEA community has neither shown the existence of such a trend, nor have they expressed their criticisms in these terms, but rather in vaguer ones. Their explicit formulation is that this approach 'could' overstate the effects of ACC because of its focus on thermodynamic variables. This merely refers to a possibility but not to a general trend, as interpreted by these scholars. However, the formulation of this criticism and the reception of their ideas by the scientific community suggest, at least, the existence of a general association between the use of the storyline approach and overstatements of ACC in the literature, presumably because of the conditional nature of this approach and its focus on thermodynamic variables.

This chapter offers arguments that cast doubt on this general association, leaving aside the question of whether the storyline approach is prone or not to overstatements. The examples below aim to highlight cases where this general association might not hold. With this point, this chapter at least shows that the conditional nature of the storyline approach does not necessarily lead to overstatements of ACC. The arguments offered here do not disprove statements about general trends (i.e., they do not show that the storyline approach does not overstate the effects of ACC) and that they do not show the opposite trend (i.e., that the storyline approach understates the effects of ACC).[10] Nonetheless, this discussion provides two important contributions in this regard. First, that if the storyline approach were prone to overstate the effects of ACC, then it would be so for reasons other than those provided by PEA proponents (i.e., not *simply* because of the conditional nature of the approach). Second, that the storyline approach can also understate the effects of ACC and they also need to be taken into consideration if one aims to analyze the process of each method toward over- or understatements.

Finally, for clarity's sake, one should differentiate between two plausible interpretations of the complaint that the storyline approach overstates the effects of ACC, which differ mostly in their respective focus.[11] First, the criticism could focus exclusively on weather event itself. According to this interpretation, the complaint of the PEA community against the storyline approach would be that the storyline approach states that ACC has increased the magnitude of the EWE more than it has in fact done it. For example, here the complaint could be that the storyline approach affirms that ACC caused extreme temperature happening in a particular geographical location at a particular time to be X degrees higher than expected from natural variability, when in fact ACC only caused that extreme temperature to be X–Y (Y>0) degrees higher. Call this interpretation of the criticism of overstatement 'O1'.

Second, the criticism could be focused on the attribution of certain harmful effects or impacts to ACC. According to this interpretation, the complaint of the PEA community would be that the storyline approach often suggests that certain negative impacts are attributable to ACC, when in fact they should not be attributed to ACC. Call this interpretation of the criticism of overstatement 'O2'. The concern that the storyline approach overstates the effects of ACC in this sense appears implicitly when the PEA community warns against the danger of overadaptation triggered by the results of storyline studies. The concern of the PEA community is that the storyline approach might exaggerate the impacts occurring in particular location due to ACC, thereby suggesting investing money to adapt to the negative effects of ACC where in fact is not needed (Stott et al. 2013; Stott et al. 2016; Stott, Karoly, and Zwiers 2017).

Note that these two interpretations of the PEA community's criticism are not mutually exclusive, but rather closely related. In fact, they are often implied together in the criticisms of overstatements raised by the PEA community against the storyline approach. Here, I distinguish them for analytic purposes and highlight that they differ mostly in their respective focus: whereas O1 focuses on the link between changes in climatic conditions and ACC, O2 focuses on the attribution of negative impacts to ACC when certain thresholds are overshot.[12] Depending on the context of the discussion, one or the other focus might be more relevant to reflect and understand the criticisms of the PEA community.

The next section argues that the storyline approach does not necessarily overstate the effects of ACC for the reasons offered by the PEA community. Later,

Section 5 moves on to provide an independent argument showing that the probabilistic approach is vulnerable to similar criticisms raised against the storyline approach.

5.3 The storyline approach and the criticism of overstatement

Let us start with O1. The complaint here would be that the storyline approach overstates the effects of ACC because it affirms too much of an increase in the magnitude of an EWE due to ACC. The complaint by the PEA community is based on the conditional nature of the storyline approach, that is, on the fact that the storyline approach focuses only on the thermodynamics and takes the dynamic variables as fixed. The combination of two factors might support the belief that the storyline approach would typically show an increase in the magnitude of an EWE. These factors are the well-reported general increase in global temperature due to climate change and the focus of the storyline approach on thermodynamic climate variables. The idea could be that because of the robust connection between increases in global temperature and ACC, an approach that is focused on thermodynamic changes (strongly related to temperature changes) would typically find increases in the magnitude of the event due to ACC.

Note that this complaint assumes that the dynamics of the atmosphere that are taken as a given for a certain EWE make the results of the storyline approach affirm that ACC had increased the magnitude of the event more than if we could account for dynamic changes caused by ACC, which, in fact, contributed to the EWE. Or, in other words, that complaint assumes that if we could reliably account for the actual changes in the dynamics due to ACC (which often remain uncertain), we would see that ACC made the event less severe than the storyline approach shows. Scientists working with the probabilistic approach have emphasized this point by offering case studies where detectable dynamic changes reduce the effects of ACC in comparison with these shown by only taking thermodynamic changes into consideration (Otto 2015; Otto et al. 2016; Pall et al. 2011). For instance, Otto et al. (2016) refer to the heavy flooding in Germany in 2013, where some parts of the southeast region received a month's worth of precipitation in three spring days (Schaller et al. 2016). As they argue, we would expect the likelihood of this event to increase with ACC because the vapor capacity of the atmosphere increases with warming. In fact, the study shows that an increase of 0.9 K of temperature in the region and season would increase the likelihood of such rains by approximately 6%. Such an increase would render a 1-in-200-year event in a preindustrial climate a 1-in-120-year event with ACC. However, simulations of the overall change in risk show no change in the likelihood of the event occurring. This result implies, as Otto et al conclude, that there is an important role of atmospheric circulation in counteracting the increase in probability that would be expected by only considering thermodynamic factors (Otto et al. 2016, 815).

However, this concern ignores the fact that actual changes in the dynamics could also make an EWE more severe than shown by only considering thermodynamic changes caused by ACC. In fact, cases of this sort can also be found in the literature, where detectable dynamic changes show the opposite effect, that is, an increase in the detected effects of ACC in comparison to these detected when only tracking thermodynamic changes (Schaller et al. 2016; Pfahl, O'Gorman, and Fischer 2017). In these cases, a conditional approach – such as the one used in most storyline approach studies – understates the effects of ACC because its results would only be based on thermodynamic changes. Thus, in a nutshell, the fact that the storyline approach fixes the dynamic variables and focuses on the thermodynamics does not necessarily make this approach overstate increases in the magnitude of EWEs. Hence, it is not true that by focusing only on the thermodynamics, the storyline approach necessarily overstates the effects of ACC according to O1.

Let us now turn to O2. To recall, according to this interpretation, the complaint of the PEA community would be that the storyline approach suggests that ACC has caused (or will cause) certain harmful effects when in fact it has not. Following the PEA community's concerns, it is worth exploring whether the storyline approach necessarily overstates the effects of ACC in this sense, thereby suggesting, for instance, to invest adaptation funds in regions that are in fact not affected by climate change (not at least to a significant degree), or to make polluters liable for the harmful effects of ACC in a certain region.

First, it is worth recalling that studies conducted with the storyline approach do not only report increases but also decreases in the magnitude of EWE (see Section 5.1). Since most harmful impacts are caused by increases in the magnitude of EWE, the storyline approach would not suggest that ACC has caused certain negative effects when reporting decreases in the magnitude of certain EWE. This remark rejects the idea that the storyline approach always implies that ACC has an impact on harmful effects caused by local weather events, as implied by Allen (2011).

Nevertheless, here, the combination of the following factors might again support the belief that the storyline approach would typically suggest that the harmful impacts associated with an EWE have been caused by ACC, or at least that ACC has significantly contributed to causing those impacts. As before, two factors are the well-reported general increase in global temperature due to climate change and the focus of the storyline approach on thermodynamic climate variables. The third one is the positive effect of temperature increases on many hazards (hotter heatwaves, more intense rain, etc.). The idea here would be that because of the robust connection between increases in global temperature and ACC and the connection between temperature increases and many hazards, an approach that is focused on thermodynamic changes (strongly related to temperature changes) would identify increases in the magnitude of a local event due to ACC and thus a connection between ACC and certain harmful impacts. This again might suggest that the storyline approach always finds that ACC has caused or at least has had a significant contribution to the occurrence of harmful impacts, thereby overstating the effects of ACC.

Second, however, as mentioned earlier, recall that accounting for the effects of ACC in dynamic changes could also show that in fact the EWE was made more severe than shown by only considering thermodynamic changes caused by ACC (i.e., with storyline approach studies). In this case, only considering thermodynamic changes would underestimate the magnitude of the impacts attributable to ACC. Accordingly, then, storyline approach studies could also be underestimating the harmful effects caused by ACC.

Third, and perhaps most importantly, the storyline approach could be combined with a decision threshold to limit the attribution of harmful impacts to ACC, similar to the one suggested by some PEA scholars. Recall that PEA studies do not attribute all EWE to ACC for which some increase in probability is detected. Instead, scholars working with the probabilistic approach have suggested that PEA studies could use a threshold (usually, 0.5 FAR) of increased likelihood to attribute EWEs to ACC. The idea is that when this threshold is exceeded, following the standards of certain legal contexts, one could attribute the EWE to ACC. Often, PEA scientist imply that this procedure could not only be used to attribute the EWE themselves but also their harmful impacts to ACC. This is made especially clear when suggesting that this procedure could be used in liability and compensatory contexts (Allen 2003; Allen and Lord 2004; Allen et al. 2007; Allen 2012; Thompson and Otto 2015). One cannot seek compensation or be made liable for the occurrence of extreme rainfall precipitation or for extreme temperatures alone, but rather for their negative impacts on their property or their health. Hence, the claim that attribution studies linking EWE to ACC could be used in liability and compensatory contexts seems to assume that the decision threshold of 0.5 FAR not only serves the purposes of attributing the EWE to ACC but also to attribute the harmful impacts resulting from that EWE to ACC. That is, the 0.5 FAR seems to work implicitly as a threshold to attribute certain negative impacts to climate change, even if the selection of that threshold is only justified on legal grounds. This threshold is not intrinsic to PEA, but it is rather an addition to decide which impacts are attributable to ACC and which are not, for legal, political, economic, or other societal purposes. Notably, this threshold implies that some of the effects of ACC are left aside or ignored. For instance, the use of this threshold would exclude the attribution of any climate change-related impacts for those EWE for which a probability increase is positive but less than 100% (i.e., 0.5 FAR), even if the probability of the event has been increased by 99% due to ACC.

The point here is that *if* something like this is acceptable for the probabilistic approach, the results of the storyline approach could also be combined with a threshold to decide whether the impacts associated with a certain EWE are so relevant as to be attributed to ACC for similar legal, political, economic, or other kinds of societal purposes. For instance, imagine that we want to assess whether certain impacts (for instance, a flood, or the property losses derived thereof) associated with an EWE (for instance, heavy rainfall) can be attributable to ACC, perhaps for deriving compensatory claims. In this case, one would not refer to a ratio of conditional probability, as in PEA studies. But, instead, one could use an absolute meteorological threshold to derive impact attribution claims, which would exclude some of the results derived from storyline approach studies. That is, one might artificially use this threshold to derive attribution claims for the impacts of EWE that reach a certain magnitude and exclude those that do not reach this magnitude. Such a relevant meteorological threshold for the storyline approach could be derived from expert knowledge or from impact model studies assessing the sensitivity of an impact to climate change.[13] For instance, in case of a flood caused by heavy precipitation, the meteorological threshold could be informed by a relevant increase in runoff or by a relevant increase of the damage to some critical infrastructure. Also, a threshold could be defined relative to natural fluctuations of an impact, for example, typical variations in runoff. If the identified effect would exceed a chosen multiple of these fluctuations (e.g., 2σ), this effect would be attributed to ACC. Such a threshold would be conceptually similar as the PEA threshold (both compare a signal with natural fluctuations), but it would not be identical in the sense that it would not necessarily issue the same attribution statements as the PEA approach. In any case, the point here is not to provide a unique approach to define such a threshold but to highlight that different possible avenues could be used to define a threshold for attribution statements that can serve different purposes.

Even if we believed that the focus of the storyline approach on thermodynamic changes tends to overstate the effects of ACC in the sense O1 (i.e., by overstating the magnitude of the weather event itself), a decision threshold similar to the one implied or suggested by some PEA scientists could prevent, or at least limit, overstatements in the sense O2 when using the storyline approach (i.e., overstatements of impacts due to ACC).[14]

Here, two important points deserve attention. First, this decision threshold would be an artificial addition to the results obtained from the storyline approach, but not part of the storyline studies themselves. But this should not be a reason to reject this possibility *if* one accepts the relevance of this (similar) procedure for PEA studies. Second, there is no reason to believe that this procedure would overstate the effects of ACC in the sense of overstating its harmful impacts (overstatements in sense O2) if the storyline approach is employed more than if the probabilistic approach is employed. In fact, whether this kind of impact attribution would overstate the effects of ACC would depend on how thresholds are set for deciding when certain impacts are attributable to ACC for different societal purposes. This last remark should not be surprising since, as suggested by Allen and his colleagues, causal attribution claims are not only scientific issues (Allen et al. 2007, 1354). Scientific research needs to be combined with a certain understanding of causation and on a certain understanding of which thresholds are relevant to claim causation in different contexts and with different purposes.

5.4 On how the probabilistic approach is affected by similar objections

This section provides an independent argument for why the PEA criticisms do not constitute good reasons to prefer the probabilistic approach over the storyline approach. I argue that, very often, the probabilistic approach faces similar objections to those raised by the PEA community against the storyline approach. This is due to the lack of robustness of climate model simulated changes, the way events are commonly defined when applying the probabilistic approach and the stronger signal of thermodynamic changes over dynamic ones. The fact that the probabilistic approach faces similar criticisms undermines a general preference for the probabilistic approach over the alternative one, independently of the success of the previous arguments.

Let us start with the role of lack of robustness of simulated dynamic changes in climate models. As we saw, climate model results are very uncertain with respect to changes in dynamic factors. For several cases and regions, different climate models even simulate opposite *changes* in aspects related to the atmospheric circulation (Doblas-Reyes et al. 2021; Zappa, Ceppi, and Shepherd 2021) such as changes in European wind speed (Zappa and Shepherd 2017) or in central European precipitation patterns (Maraun 2013). A plausible uncertainty range can thus only be derived by an ensemble of multiple climate models that spans all plausible changes in the atmospheric circulation. State-of-the-art multimodel ensembles typically comprise some 10–30 different (although not independent, as they may share several components) climate models (e.g., Eyring et al. 2016; Jacob et al. 2016). But because of the high computational costs of simulating climatic changes under different forcings, many classical event attribution studies have been conducted based on a single model only (e.g., Stott et al. 2004; Pall et al. 2011; Lott, Christidis, and Stott 2013). The world weather attribution (WWA) initiative demands 'at least two and preferably more models to be good enough for the attribution analysis' (WWA 2021). But selecting only a few climate models to represent changes in the atmospheric circulation and related phenomena can cause substantially misleading conclusions, including overstatements of the role of anthropogenic forcing.

Imagine the attribution of a heavy rainfall event in the presence of strong dynamic uncertainties. Some models may, for the considered region, simulate an increase in heavy precipitation under ACC, some a negligible change, and some a decrease. Models from the first group would suggest an increase in the likelihood and/or FAR, those from the second group an essentially constant value, and those from the third group a decrease in the likelihood and/or FAR. Given the uncertainties in dynamic changes, the true change under ACC is not known and may be in either group (or even outside, if uncertainties are not reliably sampled because of common model errors).

Selecting only a small number of models increases the danger of missing one of the groups, thereby missing the true climate change signal, and ultimately of producing an overstatement.[15] If the true effect of ACC would be an increase in the occurrence probability of the event, but the selected ensemble would only include models showing no or a negative change, the influence of ACC would be understated. But vice versa, if the true effect of ACC would be a decrease in the occurrence probability, but the ensemble would include only models showing no or positive changes, the influence of ACC would be overstated.

The problem is aggravated by the criteria recommended for selecting suitable models for event attribution (Mitchell et al. 2017; WWA 2021; van Oldenborgh et al. 2021): they are all based on the performance at reproducing key aspects of extreme events in the *present climate*, but this does not ensure a credible representation of *changes* in extreme events. Let us assume that the spread in climate change signals across the full ensemble would represent the true uncertainty we have about climate change.[16] Selecting only a subset of these models – and thus reducing ensemble spread – without giving any physical argument of why this subsampling should reduce the true uncertainty (i.e., with an argument that links present-day model performance to the credibility of the climate change signal) would thus in general underestimate the true uncertainties. To fully represent uncertainties, an additional model selection criterion regarding the representation of model spread is thus required, but this is not included among the listed criteria.[17] In any case, the key point is: if only a small number of models are considered, the probabilistic approach can, depending on the case and model choice, yield overstatements of the role of ACC.[18]

Let us now turn to the role of defining the event under consideration. Recall the PEA complaint that the storyline approach overstates the effects of ACC by only considering the thermodynamics. Presumably, the PEA community takes this to be a reason to favor the probabilistic approach because, in theory, this approach considers both the dynamic and thermodynamic variables. However, as it will be shown, the way events are commonly defined in PEA studies focus very often only on the thermodynamic variables of the EWE, leaving aside the dynamic ones.

Recall that the probabilistic approach asks about the occurrence of type of events. For conducting their counterfactual analysis, PEA studies must define the kind of event they are interested in with some level of abstraction. In doing so, PEA studies operate with a simple, one-dimensional definition of the event. This simple definition leaves aside some of the atmospheric, meteorological factors and temporal aspects characterizing the particular event that motivated the attribution question, which is essentially multidimensional. The one dimensional definitions are called here 'proxy-definitions', as they constitute only a simple approximation to the set of atmospheric and meteorological conditions characterizing the particular event that led to impacts raising the public's interest.

The point here is that these proxy-definitions are often designed in a way that leaves aside or downplays dynamic factors, as it occurs in studies conducted with the storyline approach. For instance, extreme events such as droughts and heatwaves are often caused by an interplay of dynamic and thermodynamic aspects. In the mid-latitudes, dry spells and heatwaves are typically caused by persistent blocking high-pressure systems (Woollings et al. 2018). However, PEA attribution studies typically express these multidimensional events, which have a distinct dynamic aspect, with simple, one-dimensional definitions. For instance, the 2003 European heatwave was caused by a blocking high, persistent for several weeks and amplified by low-soil-moisture conditions (Fischer et al. 2007). But, in their attribution study of the 2003 European heatwave, Stott et al. (2004) characterized the event by *the June-August mean temperature*. Similarly, in 2018 Western, Central and Northern Europe were struck by a severe several-month-long drought caused by recurring blocking conditions and accompanied by several heatwaves (WMO 2019). But the World Weather Attribution initiative characterized this event by the *3-day maximum temperature average in 2018* (WWA 2018). In both event definitions, the dynamic state – recurring persistent blocking – is ignored.

Admittedly, such one-dimensional, proxy-definitions, focused on thermodynamic factors, might have several advantages. First, state-of-the-art climate models still have substantial limitations in representing the dynamics underlying such events, in particular the persistence (Weisheimer, Palmer, and Doblas-Reyes 2011; Mitchell et al. 2017; Schiemann et al. 2020), and averaging across longer time scales, or selecting a short period helps to navigate these limitations as seasonal means and daily statistics are usually well represented. Second, similarly, pure temperature indices – by definition – have a strong thermodynamic climate change signal, such that dynamic uncertainties and internal variability as discussed earlier are relatively low (Shepherd 2014). Third, a one-dimensional event expressed by one number is more manageable within the standard FAR framework.

However, these proxy-definitions do not capture the dynamic aspects of the event. This is important for two reasons. Firstly, although focusing on the thermodynamic aspects of the event might have those advantages, it might lead to interpretations that overstate the effects of ACC in both sense O1 (overstatements focused on the weather event) and O2 (overstatements focused on the impacts).

Consider the example of heatwaves, which include a temporal aspect (dynamics: the persistence of the blocking high) and the actual temperatures (thermodynamics). Climate models simulate a broad range of plausible changes in the frequency of European summer blocking yet with an overall tendency toward fewer events (Davini and D'Andrea 2020). But all climate models simulate a robust increase in European summer mean temperatures and 3-day maximum temperatures (Gutiérrez et al. 2021). Thus, even though the kind of multidimensional events that triggered the attribution question (the 2003 heatwave and the 2018 and 2022 droughts, which included a dynamical aspect) could potentially become less frequent in a future climate (because of the tendency toward fewer blocking events and the effects of such events on heatwaves), the simplified, temperature-based version of the events captured with proxy-definitions will become more frequent (because of robust increases in mean summer temperature or three-day summer temperature).

Similarly, the Central European Summer of 2022 was characterized by high temperatures and low rainfall with far reaching impacts on health, energy, agriculture, and municipal water supply (WWA 2022). This event had a thermodynamic component (the high temperatures) and a dynamic one (absence of rain).[19] The WWA attribution study chose summer soil moisture as an indicator to define the drought and found a clear increase in the associated FAR. But this definition highlights again the thermodynamically driven aspects of the event because reductions in soil moisture in Central Europe, particularly during the summer, are driven by evapotranspiration due to increasing temperatures (Douville et al 2021). Projected changes in the length of dry periods (i.e., the dynamically driven element of the event), which are not robust (Gutiérrez et al. 2021), are thus left aside in this definition. Again, the chosen indicator focuses on the clear thermodynamic component and downplays the role of dynamics.

Note that this procedure, in itself, does not overstate increases in the frequency of the weather event due to ACC. In fact, it is true that the simplified, temperature-based weather event captured with the proxy-definition will become more frequent because mean summer temperature and three-day maximum temperature average increase in a world with ACC. But, because the initial attribution question is raised by referring to the particular event, which is multidimensional, and the attribution study is carried out for a one-dimensional event captured by the proxy-definition, this increases the danger of interpreting statements about the proxy event as statements about the multidimensional event.[20] Given that those proxy events most certainly increase in a world with ACC but the multidimensional events do not necessarily do so (as described earlier), this can easily suggest that the effects of ACC on the EWE (including all its dimensions) are higher than they actually are, thereby leading to overstatements of the frequency of occurrence of certain multidimensional weather events.

Furthermore, the results of PEA might suggest or favor the interpretation that the negative impacts occurring in the aftermath of the 2003 European heatwave were only due to high mean summer temperature (thermodynamic factors), when in fact they were caused by the interplay between thermodynamic and dynamic factors. Such an interpretation might overstate the impacts associated with ACC because ACC increases mean summer temperatures but not necessarily the set of atmospheric and meteorological conditions that fully characterized the 2003 event and their associated impacts. For instance, the impacts of the 2003 event were also influenced by the blocking high- and low-soil-moisture conditions. Ignoring the influence of those conditions, which are more uncertain because they are closely tied to the dynamic aspect of the event, might suggest an overstatement of the impacts associated with ACC. In fact, there are reasons to believe that ACC might decrease the frequency of that set of conditions, given that it decreases blocking events present in the 2003 event. For this reason, the probabilistic approach is not free from objections similar to those raised against the storyline approach, which undermines the preference for the former over the latter on those grounds.

Secondly, and most importantly, with this use of proxy-definitions, note that PEA studies operate in a very similar way to the storyline approach. Ironically, by using a temperature-based proxy-definition, the probabilistic approach essentially disregards dynamic changes and emphasizes the thermodynamic ones. Arguably, this practice is not much different from the one characterizing the storyline approach (i.e., the practice of conditioning on an unchanged atmospheric circulation and focusing on thermodynamic changes).[21]

The claim is not that the use of proxy-definitions (focused on the thermodynamics) always and necessarily affects PEA studies. In principle, the same type of problem may arise for the other types of events considered by WWA, in particular when these occur on time scales not well represented by climate models. It is not further explored here whether and how this argument could be extended to other events. However, this argument works at least for the events listed here.

Finally, there is a further way in which PEA studies might focus on thermodynamic changes. This point concerns the stronger signal of thermodynamic changes, which might outweigh the role of dynamic changes. This point might affect all event types considered by the WWA, but here a flood event serves to illustrate this point.

Consider the severe flooding in Germany and Belgium in July 2021. A slowly moving cut-off low caused unprecedented amounts of rainfall, corresponding to a 400-year event in present climate (Tradowsky et al. 2023). According to a PEA study, climate change has increased the occurrence probability of such an event by a factor of 1.2 to 9; that is, the observed rainfall amounts would have been extremely unlikely without climate change (Tradowsky et al. 2021). This attribution statement is based on the overall changes in rainfall, which themselves are caused by the combined effect of changes in the occurrence of cut-off lows (dynamics) and in the rainfall intensities within cut-off lows (thermodynamics). To understand why this statement focuses on thermodynamic changes, we need to assess both changes separately.

Although currently there are no analyses of cut-off low changes in climate simulations, it has been suggested to use changes in blocking highs as a proxy for cut-off low changes (Maraun et al. 2022), given that cut-off lows tend to develop along with blocking highs (Nieto et al. 2007). Current generation climate models show a large spread of changes in the number of days with a blocking high due to ACC, ranging from increases to decreases, but with a slight decrease when considering the mean over all models (Davini and D'Andrea 2020). These uncertainties arise from both model uncertainty and internal climate variability (see Woollings 2010; Woollings et al. 2018). Transferring this finding to cut-off lows, we have substantial uncertainty about the influence of climate change on their occurrence but expect a slight decrease. But, as sketched earlier, despite this uncertainty and the overall decrease in event occurrence, the overall PEA states an increase in the occurrence probability of the observed heavy rainfall.

The most plausible explanation of this seemingly contradictory result is a strong increase in the rainfall intensities given a cut-off low, which outweighs the decrease in event occurrence and the large uncertainties about this decrease. In other words, even though the PEA considers the full attribution including changes in dynamics and thermodynamics, it mostly draws its strength from thermodynamic changes.[22] This argument could be easily transferred to the other types of events considered by WWA.

To close this section, recall that defenders of the probabilistic approach believe that the practice of ignoring dynamic changes is a weakness of the storyline approach and that, at least implicitly, this constitutes a reason to prefer their own approach. However, if PEA operates in a similar manner, quite straightforwardly, the fact that the storyline approach focuses only on thermodynamic variables cannot be a reason to disregard the storyline approach in favor of the probabilistic approach.[23]

5.5 Conclusion

Attribution science is evolving rapidly, and the emergence of new alternative methods triggers the question of which method to follow when it comes to attributing

EWE to ACC. Different variables might be relevant to decide on this matter. Among them, there is the performance of each attribution method in estimating the effects of ACC. If we had reasons to believe that one method overstates the effects of ACC, whereas the other does not, this might give us reasons to prefer one method over the other, particularly in liability for loss and damage contexts. For that reason, it is worth investigating the association, suggested by the PEA community, of the storyline approach with overstatements of ACC and also whether the criticisms offered against the storyline approach constitute good reasons to prefer, in general, the probabilistic approach. In this chapter, I have argued that there are reasons against such a general association and that the probabilistic approach faces, at least sometimes, similar objections to those pressed by its proponents against the storyline approach.

First, I have argued that the storyline approach does not always overstate the effects of climate change. In a nutshell, I have argued that the fact that the storyline approach fixes the dynamic variables and focuses on the thermodynamics does not make this approach inherently likely to overstate increases in the magnitude of EWEs because unknown dynamic changes could have also made the EWE more severe than shown by the storyline approach. Moreover, I argued that this approach does not necessarily overstate the harmful impacts that are attributable to ACC because this depends on how certain thresholds are established, as it occurs with the probabilistic approach.

Second, independently, the probabilistic approach faces similar objections to those raised by the PEA community against the storyline approach. The lack of robustness of climate models might, in many circumstances, make the results of the probabilistic approach overstate the effects of ACC, depending on the model selection. Moreover, the use of temperature-based proxy-definitions might lead to interpretations of the results provided by PEA studies that might overstate the role of ACC on specific EWEs. Furthermore, and most importantly, proxy-definitions essentially deemphasize the dynamic components, thereby operating in the way the PEA community criticized the storyline approach. Finally, something similar might happen when thermodynamic changes dominate over dynamic ones. Thus, the fact that the probabilistic approach faces similar criticisms does not justify a general preference for this approach over the alternative one, independently of the success of the previous arguments.

Notes

1 Additionally, some have claimed that attribution studies are relevant for adaptation measures because 'based on the occurrence of a particularly damaging extreme event, plans could be made to adapt to an increasing frequency of such events in future' (Stott et al. 2016, 24; similarly in Stott et.al. 2017).

2 One can often find the term 'probabilistic approach' in the literature to refer to this approach. However, this term is here avoided because it implicitly conveys the misleading idea that other approaches (i.e., the storyline approach) do not provide a risk-assessment.

3 Similarly, those decisions can also influence adaptation cases. If stakeholders need to decide between allocating adaptation funds to a particular region or investing these funds on development aid in a different one, they might want to avoid methods that overstate the effects of ACC. After all, the results of such a method would presumably make the adaptation problem look worse than it is and therefore generate some biases toward the adaptation project.

4 Here two things should be noted. First, exceeding this threshold means that ACC has increased the probability of the event by more than 100% or has more than doubled the risk of the EWE occurring. Second, here that the suggestion of using the FAR value to derive attribution claims goes beyond simply attributing certain climatic conditions to ACC. Instead, the focus here is attributing certain harmful impacts to climate change. These are the focus of the PEA community when they suggest using this approach for justice or legal purposes (Allen 2003; Allen et al. 2007; Allen 2012; Thompson and Otto 2015; Otto et al. 2017). Also, see the discussion below.

5 The storyline approach is not specifically designed for event attribution. Physical climate storylines have been defined as a self-consistent and plausible physical trajectories of the climate system, or a weather or climate event, on time scales from hours to multiple decades (Shepherd et al. 2018; Chen et al. 2021, section 1.4.4.2) Such physical trajectories describe plausible future scenarios or past events and therefore can serve various purposes, from risk assessment of plausible climate change–related impacts to attribution of EWEs (see IPCC report, chapter 11, section 2.3; Sillmann et al. 2021). Extreme event attribution relies on storylines of observed and counterfactual events. Moreover, note that the scientific community does not see physical climate storylines as a single concept (Jack et al. 2020).

6 A longer discussion is provided in 5.4. Moreover, this is the only problem that PEA studies might face or that has been identified by other scientists. Others might include, for instance, the lack of a long enough observational record (van Oldenborgh et al. 2021). However, an extended discussion of these issues is beyond the scope of this chapter.

7 Even more, scientists have started to work on how the storyline approach could be applied also to legal contexts. See Lloyd and Shepherd (2021).

8 Note that overstating and making false positives are not coextensive concepts. One can overstate an effect without necessarily claiming a false positive because no null hypothesis is tested, and not all false positives are overstatements because that depends on how the null hypothesis had been formulated. However, for the sake of the discussion, the term overstating is used here because it captures the concerns behind this controversy in a more overarching way.

9 More details of the controversy have recently been reported in various pieces by Elisabeth Lloyd, Naomi Oreskes and Eric Winsberg (Lloyd and Oreskes 2018, 2019; Winsberg, Oreskes, and Lloyd 2020).

10 This point was brought up by an anonymous reviewer.

11 Arguably, these do not exhaust all possible forms of overstatement. For instance, an attribution method could also overstate the effects of climate change by overstating decreases in likelihood or magnitude of an EWE. In this sense, an attribution method would affirm too much of a decrease in any of these parameters. However, this interpretation of overstatement is unlikely to capture the concern raised by the PEA community. Arguably, the reason is that overstating a decrease in likelihood or magnitude rarely implies overstating harmful impacts because these are mostly driven by increases (not decreases) in the likelihood and magnitude of EWE. For instance, cold spells are among the EWEs that have likely decreased in likelihood and magnitude due to ACC. However, decreases in cold spells rarely cause harm because societies tend to be adapted to average weather patterns. A decrease in the likelihood and magnitude of cold spells means that local weather moves closer to the average and thus stays more stable and within

the limits of adaptation. This interpretation is left aside because the harmful impacts of climate change are the underlying focus of many attribution papers (Allen 2003; Allen et al. 2007; Allen 2012; Thompson and Otto 2015; Otto et al. 2017).

12 O1 and O2 are closely connected but their relation is not necessarily of implication. First, overstatements of the magnitude of an EWE that only refer to how climatic conditions have been affected by ACC (O1) might not imply overstatements of the impacts caused by climate change. The reason is that an increase in the magnitude of an EWE (say, extreme temperature) might not be significant enough to overshoot thresholds associated with the occurrence of certain impacts, given that the relation is often not linear. However, probabilistically, overstatements in sense O1 lead to overstatements in sense O2. Second, O2 often implies O1 because, in this context, overstatements of the impacts attributable to ACC (O2) are derived precisely from overstatements of the magnitude of the EWE because of ACC (O1). However, it is also possible that the severity of an event is not overstated but that the impacts resulting from the occurrence of an event of this severity are overstated. For instance, one might accurately describe the effects of ACC on rainfall (event) but overstate the impacts of rainfall on landslides, particularly if one misrepresents the effects of the interplay between rainfall and other Causation Objection-founding factors (such as soil moisture) on the severity and occurrence of landslides (Perkins-Kirkpatrick et al. 2022).

13 Such studies derive so-called response surfaces or impact functions (e.g. Prudhomme et al. 2010), which quantify the response of an impact (e.g., river runoff) to changes in key drivers.

14 Admittedly, such a threshold would not prevent all possible overstatements. In cases in which the storyline overstates how much ACC increased the severity of events whose magnitude is situated above the threshold to attributing certain harmful effects to ACC, this methodology would indeed overstate the effects of ACC in sense O2. However, this threshold would at least limit the risk of overstating the impacts of ACC. Furthermore, it is worth noting that the combination of PEA with a 0.5 FAR threshold is also not free from threshold-related problems since it delivers the same attribution statements for, for example, 0.6 FAR events and 1.0 FAR events. That is, the (arguably important) differences in changes on the probability of occurrence due to ACC are not reflected in the attribution statements.

15 I am aware that selecting the full ensemble across all groups will avoid overstatements of ACC because opposing changes may result in an inconclusive statement about the influence of ACC ('we don't know yet') (Shepherd 2016). Probably, this is the reason why it has been argued that the probabilistic approach tends to understate the effects of ACC (Winsberg, Oreskes, and Lloyd 2020; Lloyd and Oreskes 2019). However, the (not uncommon) selection of a small number of models can also yield overstatements, something that has been underemphasized in the literature.

16 In general, model ensembles are not designed to fully sample uncertainties but based on availability. They are referred to as ensembles of opportunity and typically underestimate true uncertainties (Tebaldi and Knutti 2007).

17 Selecting models to reduce climate change uncertainties is a challenging topic of active research (Eyring et al. 2019) but essentially unresolved. Thus, the IPCC states:

> there is high confidence that ensembles for regional climate projections should be selected such that models unrealistically simulating processes relevant for a given application are discarded, but at the same time, the chosen ensemble spans an appropriate range of projection uncertainties.
>
> (Doblas-Reyes et al. 2021)

18 Although I am not aware of case studies showing how the use of only a few models yields overstatements, these argument hints at this possibility, for the reasons provided in the main text. Partly in response to this problem, the approach to estimating the

FAR has been generalized to be able to make use of existing multimodel ensembles (WWA 2021; van Oldenborgh et al. 2021), although the recommendations of the WWA for model selection should be amended by a criterion that includes representing the full spread in relevant dynamic aspects. However, in practice, many studies are still based on a small number of models. Some examples include Otto et al. (2018) and Kirchmeier-Young et al. (2017)

19 Note that presence/absence of rain is determined by the weather type (meteorological drought) and thus by the dynamics. The intensity or severity of the rain would be related to the thermodynamics (Clausius Clapeyron).

20 As a matter of example, Stott et al.'s (2004) paper seemingly aims to analyze the 'human contribution to the European heatwave of 2003' (as the title goes). Hence, the initial and motivating attribution question refers to the 2003 event, which was caused by the interplay of various dynamic and thermodynamic factors. However, the results of the PEA study only show the influence of ACC on 'unusually high mean summer temperatures'. The point here is that if the results of this PEA study, for instance, are interpreted as providing an answer to the initial attribution question concerning the multidimensional 2003 event (which would not be surprising given that the results are provided as an answer to such a question), such an interpretation might lead to overstatements. I am aware that this is not an inherent problem of PEA studies but rather on how they are received by stakeholders. However, the risk of misinterpretation and thus of overstatements is real, and it deserves attention, even if it is only to avoid them.

21 Some researchers have developed attribution approaches for compound events (Mazdiyasni et al. 2019; Kiriliouk and Naveau 2020), which can account for the dynamic aspects (and other variables such as precipitation or wind; Zscheischler et al. 2020), but these approaches are not widely used and still suffer from the lack of robustness in projecting dynamical changes.

22 Note that this argument holds for the mid-latitudes. In the subtropics, also dynamic changes may be very robust because of their direct link to the Hadley cell (Cresswell-Clay et al. 2022).

23 Note that the use of proxy-definitions also challenges the idea that the probabilistic approach treats an EWE as a 'single, self-reinforcing and indivisible whole' (Allen 2012, 13); see Section 5.1.

References

Allen, Myles. 2003. 'Liability for Climate Change'. *Nature* 421: 891–92

———. 2011. 'In Defense of the Traditional Null Hypothesis: Remarks on the Trenberth and Curry WIREs Opinion Articles'. *Wiley Interdisciplinary Reviews: Climate Change* 2 (6): 931–34.

———. 2012. 'The Scientific Basis for Climate Change Liability'. In *Climate Change Liability: Transnational Law and Practice*, edited by Richard Lord, 8–22. Cambridge; New York: Cambridge University Press.

Allen, Myles, Pardeep Pall, Daithi Stone, Peter Stott, David Frame, Seung-Ki Min, Toru Nozawa, and Seiji Yukimoto. 2007. 'Scientific Challenges in the Attribution of Harm to Human Influence on Climate'. *University of Pennsylvania Law Review*, 1353–400.

Allen, Myles R., and Richard Lord. 2004. 'The Blame Game: Who Will Pay for the Damaging Consequences of Climate Change'? *Nature* 432.

Burger, Michael, Jessica Wentz, and Radley Horton. 2020. 'The Law and Science of Climate Change Attribution'. *Columbia Journal of Environmental Law* 45 (1).

Chen, D., Rojas, M., Samset, B. H., Cobb, K., Diongue Niang, A., Edwards, P., Emori, S., Faria, S. H., Hawkins, E., Hope, P., Huybrechts, P., Meinshausen, M., Mustafa, S. K., Plattner, G.-K., & Tréguier, A.-M. (2021). Framing, context, and methods. In *Climate Change 2021: The Physical Science Basis. Contribution of Working Group I to the Sixth*

Assessment Report of the Intergovernmental Panel on Climate Change, edited by V. Masson-Delmotte, P. Zhai, A. Pirani, S. L. Connors, C. Péan, S. Berger, N. Caud, Y. Chen, L. Goldfarb, M. I. Gomis, M. Huang, K. Leitzell, E. Lonnoy, J. B. R. Matthews, T. K. Maycock, T. Waterfeld, O. Yelekçi, R. Yu, & B. Zhou, 147–286. Cambridge University Press. https://doi.org/10.1017/9781009157896.003

Cresswell-Clay, Nathaniel, Caroline C. Ummenhofer, Diana L. Thatcher, Alan D. Wanamaker, Rhawn F. Denniston, Yemane Asmerom, and Victor J. Polyak. 2022. 'Twentieth-Century Azores High Expansion Unprecedented in the Past 1,200 Years'. *Nature Geoscience* 15 (7): 548–53. https://doi.org/10.1038/s41561-022-00971-w.

Curry, Judith. 2011. 'Nullifying the Climate Null Hypothesis'. *Wiley Interdisciplinary Reviews: Climate Change* 2 (6): 919–24. https://doi.org/10.1002/wcc.141.

Davini, Paolo, and Fabio D'Andrea. 2020. 'From CMIP3 to CMIP6: Northern Hemisphere Atmospheric Blocking Simulation in Present and Future Climate'. *Journal of Climate* 33 (23): 10021–38. https://doi.org/10.1175/JCLI-D-19-0862.1.

Doblas-Reyes, F. J., A. A. Sorensson, M. Almazroui, A. Dosio, W. J. Gutowski, R. Haarsma, R. Hamdi, B. Hewitso, W.-T. Kwon, B. L. Lamptey, D. Maraun, T. S. Stephenson, I. Takayabu, L. Terray, A. Turner, and Z. Zuo. 2021. 'Linking Global to Regional Climate Change'. In *Climate Change 2021: The Physical Science Basis. Contribution of Working Group I to the Sixth Assessment Report of the Intergovernmental Panel on Climate Change*. Cambridge: Cambridge University Press.

Douville, H., K. Raghavan, J. Renwick, R. P. Allan, P. A. Arias, M. Barlow, R. Cerezo-Mota, A. Cherchi, T. Y. Gan, J. Gergis, D. Jiang, A. Khan, W. Pokam Mba, D. Rosenfeld, J. Tierney, and O. Zolina. 2021. 'Water Cycle Changes. In Climate Change 2021: The Physical Science Basis'. In *Contribution of Working Group I to the Sixth Assessment Report of the Intergovernmental Panel on Climate Change*, edited by V. Masson-Delmotte, P. Zhai, A. Pirani, S. L. Connors, C. Péan, S. Berger, N. Caud, Y. Chen, L. Goldfarb, M. I. Gomis, M. Huang, K. Leitzell, E. Lonnoy, J. B. R. Matthews, T. K. Maycock, T. Waterfield, O. Yelekçi, R. Yu, and B. Zhou, 1055–210. Cambridge; New York: Cambridge University Press.

Eyring, Veronika, Sandrine Bony, Gerald A. Meehl, Catherine A. Senior, Bjorn Stevens, Ronald J. Stouffer, and Karl E. Taylor. 2016. 'Overview of the Coupled Model Intercomparison Project Phase 6 (CMIP6) Experimental Design and Organization'. *Geoscientific Model Development* 9 (5): 1937–58. https://doi.org/10.5194/gmd-9-1937-2016.

Eyring, Veronika, Peter M. Cox, Gregory M. Flato, Peter J. Gleckler, Gab Abramowitz, Peter Caldwell, William D. Collins, Bettina K. Gier, Alex D. Hall, Forrest M. Hoffman, George C. Hurtt, Alexandra Jahn, Chris D. Jones, Stephen A. Klein, John P. Krasting, Lester Kwiatkowski, Ruth Lorenz, Eric Maloney, Gerald A. Meehl, Angeline G. Pendergrass, Robert Pincus, Alex C. Ruane, Joellen L. Russell, Benjamin M. Sanderson, Benjamin D. Santer, Steven C. Sherwood, Isla R. Simpson, Ronald J. Stouffer, and Mark S. Williamson. 2019. 'Taking Climate Model Evaluation to the Next Level'. *Nature Climate Change* 9 (2): 102–10. https://doi.org/10.1038/s41558-018-0355-y.

Fischer, E. M., S. I. Seneviratne, P. L. Vidale, D. Lüthi, and C. Schär. 2007. 'Soil Moisture–Atmosphere Interactions During the 2003 European Summer Heat Wave'. *Journal of Climate* 20 (20): 5081–99. https://doi.org/10.1175/JCLI4288.1.

Grossman, D. A. 2003. 'Warming up to a Not-so-Radical Idea: Tort-Based Climate Change Litigation'. *Colombia Journal of Environmental Law* 28: 1–61.

Gutiérrez, J. M., R. G. Jones, G. T. Narisma, L. M. Alves, M. Amjad, I. V. Gorodetskaya, M. Grose, N. A. B. Klutse, S. Krakovska, J. Li, D. Martínez-Castro, L. O. Mearns, S. H. Mernild, T. Ngo-Duc, B. van den Hurk, and J.-H. Yoon. 2021. 'Atlas'. In *Climate Change 2021: The Physical Science Basis. Contribution of Working Group I to the Sixth Assessment Report of the Intergovernmental Panel on Climate Change*, edited by V. Masson-Delmotte, P. Zhai, A. Pirani, S. L. Connors, C. Péan, S. Berger, N. Caud, Y. Chen, L. Goldfarb, M. I. Gomis, M. Huang, K. Leitzell, E. Lonnoy, J. B. R. Matthews, T. K. Maycock, T. Waterfield, O. Yelekçi, R. Yu, and B. Zhou. Cambridge; New York: Cambridge University Press. In Press.

Hannart, A., J. Pearl, F. E. L. Otto, P. Naveau, and M. Ghil. 2016. 'Causal Counterfactual Theory for the Attribution of Weather and Climate-Related Events'. *Bulletin of the American Meteorological Society* 97 (1): 99–110.

IPCC. 2021. 'Climate Change 2021: The Physical Science Basis'. In *Contribution of Working Group I to the Sixth Assessment Report of the Intergovernmental Panel on Climate Change*, edited by V. Masson-Delmotte, P. Zhai, A. Pirani, S. L. Connors, C. Péan, S. Berger, N. Caud, Y. Chen, L. Goldfarb, M. I. Gomis, M. Huang, K. Leitzell, E. Lonnoy, J. B. R. Matthews, T. K. Maycock, T. Waterfield, O. Yelekçi, R. Yu, and B. Zhou. Cambridge; New York: Cambridge University Press.

Jack, Christopher David, Richard Jones, Laura Burgin, and Joseph Daron. 2020. 'Climate Risk Narratives: An Iterative Reflective Process for Co-Producing and Integrating Climate Knowledge'. *Climate Risk Management* 29 (January): 100239. https://doi.org/10.1016/j.crm.2020.100239.

Jacob, Daniel J., Alexander J. Turner, Joannes D. Maasakkers, Jianxiong Sheng, Kang Sun, Xiong Liu, Kelly Chance, Ilse Aben, Jason McKeever, and Christian Frankenberg. 2016. 'Satellite Observations of Atmospheric Methane and Their Value for Quantifying Methane Emissions'. *Atmospheric Chemistry and Physics* 16: 14371–96.

Kirchmeier-Young, M. C., F. W. Zwiers, N. P. Gillett, and Alex J. Cannon. 2017. 'Attributing Extreme Fire Risk in Western Canada to Human Emissions'. *Climatic Change* 144: 365–79. https://doi.org/10.1007/s10584-017-2030-0.

Kiriliouk, Anna, and Philippe Naveau. 2020. 'Climate Extreme Event Attribution Using Multivariate Peaks-Over-Thresholds Modeling and Counterfactual Theory'. *The Annals of Applied Statistics* 14 (3): 1342–58. https://doi.org/10.1214/20-AOAS1355.

Lloyd, Elisabeth A., and Naomi Oreskes. 2018. 'Climate Change Attribution: When Is It Appropriate to Accept New Methods'? *Earth's Future* 6 (3): 311–25.

———. 2019. 'Climate Change Attribution: When Does It Makes Sense to Add Methods'? *Epistemology and Philosophy of Science* 56 (1): 185–201.

Lloyd, Elisabeth A., Naomi Oreskes, Sonia I. Seneviratne, and Edward J. Larson. 2021. 'Climate Scientists Set the Bar of Proof Too High'. *Climatic Change* 165 (3): 55. https://doi.org/10.1007/s10584-021-03061-9.

Lloyd, Elisabeth A., and T. G. Shepherd. 2021. Climate Change Attribution and Legal Contexts: Evidence and the Role of Storylines. *Climatic Change* 167 (3): 1–13.

Lott, Fraser C., Nikolaos Christidis, and Peter A. Stott. 2013. 'Can the 2011 East African Drought Be Attributed to Human-Induced Climate Change'? *Geophysical Research Letters* 40 (6): 1177–81. https://doi.org/10.1002/grl.50235.

Maraun, Douglas. 2013. 'When Will Trends in European Mean and Heavy Daily Precipitation Emerge'? *Environmental Research Letters* 8 (1): 014004. https://doi.org/10.1088/1748-9326/8/1/014004.

Maraun, Douglas, Raphael Knevels, Aditya N. Mishra, Heimo Truhetz, Emanuele Bevacqua, Herwig Proske, Giuseppe Zappa, Alexander Brenning, Helene Petschko, Armin Schaffer, Philip Leopold and Bryony L. Puxley. 2022. 'A Severe Landslide Event in the Alpine Foreland under Possible Future Climate and Land-Use Changes'. *Communications Earth & Environment* 3 (1): 1–11. https://doi.org/10.1038/s43247-022-00408-7.

Mazdiyasni, Omid, Mojtaba Sadegh, Felicia Chiang, and Amir AghaKouchak. 2019. 'Heat Wave Intensity Duration Frequency Curve: A Multivariate Approach for Hazard and Attribution Analysis'. *Scientific Reports* 9 (1): 14117. https://doi.org/10.1038/s41598-019-50643-w.

Mitchell, Daniel, Paolo Davini, Ben Harvey, Neil Massey, Karsten Haustein, Tim Woollings, Richard Jones, Fredi Otto, Benoit Guillod, Sarah Sparrow, David Wallom, and Myles Allen. 2017. 'Assessing Mid-Latitude Dynamics in Extreme Event Attribution Systems'. *Climate Dynamics* 48 (11): 3889–901. https://doi.org/10.1007/s00382-016-3308-z.

Nature Editorial. 2018. 'Pinning Extreme Weather on Climate Change Is Now Routine and Reliable Science'. *Nature* 560 (7716): 5–5. https://doi.org/10.1038/d41586-018-05839-x.

Nieto, R., L. Gimeno, L. De la Torre, P. Ribera, D. Barriopedro, R. García-Herrera, A. Serrano, A. Gordillo, A. Redaño, and J. Lorente. 2007. Interannual Variability of Cut-off Low Systems Over the European Sector: the Role of Blocking and the Northern Hemisphere Circulation Modes. *Meteorology and Atmospheric Physics* 96: 85–101.

Oldenborgh, Geert Jan van, Karin van der Wiel, Sarah Kew, Sjoukje Philip, Friederike Otto, Robert Vautard, Andrew King, Fraser Lott, Julie Arrighi, Roop Singh, and Maarten van Aalst. 2021. 'Pathways and Pitfalls in Extreme Event Attribution'. *Climatic Change* 166 (1): 13. https://doi.org/10.1007/s10584-021-03071-7.

Otto, Friederike E. L. 2015. 'Attribution of Extreme Weather'. *Nature Geoscience* 8 (8): 581–82. https://doi.org/10.1038/ngeo2484.

Otto, Friederike. E. L., M. R. Allen, P. A. Stott, G. J. van Oldenborgh, J. Eden, and D. J. Karoly. 2016. 'Framing the Question of Attribution of Extreme Weather Events'. *Nature Climate Change* 6. https://ora.ox.ac.uk/objects/uuid:bd8d2c5d-17a1-4586-b545-83826f3c38a7.

Otto, Friederike E. L., Ragnhild B. Skeie, Jan S. Fuglestvedt, Terje Berntsen, and Myles R. Allen. 2017. 'Assigning Historic Responsibility for Extreme Weather Events'. *Nature Climate Change* 7 (11): 757–59, nclimate3419. https://doi.org/10.1038/nclimate3419.

Otto, Friederike E. L., Karin van der Wiel, Geert Jan van Oldenborgh, Sjoukje Philip, Sarah F. Kew, Peter Uhe, and Heidi Cullen. 2018. 'Climate Change Increases the Probability of Heavy Rains in Northern England/Southern Scotland like Those of Storm Desmond – a Real-Time Event Attribution Revisited'. *Environmental Research Letters* 13 (2): 024006. https://doi.org/10.1088/1748-9326/aa9663.

Page, Edward A., and Clare Heyward. 2016. 'Compensating for Climate Change Loss and Damage'. *Political Studies* 65 (2): 356–72.

Pall, Pardeep, Tolu Aina, Dáithí A. Stone, Peter A. Stott, Toru Nozawa, Arno G. J. Hilberts, Dag Lohmann, and Myles R. Allen. 2011. 'Anthropogenic Greenhouse Gas Contribution to Flood Risk in England and Wales in Autumn 2000'. *Nature* 470 (7334): 382–85.

Pall, Pardeep, Christina M. Patricola, Michael F. Wehner, Dáithí A. Stone, Christopher J. Paciorek, and William D. Collins. 2017. 'Diagnosing Conditional Anthropogenic Contributions to Heavy Colorado Rainfall in September 2013'. *Weather and Climate Extremes* 17 (September): 1–6. https://doi.org/10.1016/j.wace.2017.03.004.

Patricola, Christina M., and Michael F. Wehner. 2018. 'Anthropogenic Influences on Major Tropical Cyclone Events'. *Nature* 563 (7731): 339–46. https://doi.org/10.1038/s41586-018-0673-2.

Perkins-Kirkpatrick, S. E., D. A. Stone, D. M. Mitchell, S. Rosier, A. D. King, Y. T. E. Lo, J. Pastor-Paz, D. Frame, and M. Wehner. 2022 'On the Attribution of the Impacts of Extreme Weather Events to Anthropogenic Climate Change'. *Environmental Research Letters* 17 (2): 024009. https://doi.org/10.1088/1748-9326/ac44c8.

Pfahl, S., P. A. O'Gorman, and E. M. Fischer. 2017. 'Understanding the Regional Pattern of Projected Future Changes in Extreme Precipitation'. *Nature Climate Change* 7 (6): 423–27.

Prudhomme, C., R. L. Wilby, S. Crooks, A. L. Kay, and N. S. Reynard. 2010. 'Scenario-Neutral Approach to Climate Change Impact Studies: Application to Flood Risk'. *Journal of Hydrology* 390 (3): 198–209. https://doi.org/10.1016/j.jhydrol.2010.06.043.

Pulkkinen, Karoliina, Sabine Undorf, Frida Bender, Per Wikman-Svahn, Francisco Doblas-Reyes, Clare Flynn, Gabriele C. Hegerl, Aiden Jönsson, Gah-Kai Leung, Joe Roussos, Theodore G. Shepherd, and Erica Thompson. 2022. 'The Value of Values in Climate Science'. *Nature Climate Change* 12 (1): 4–6. https://doi.org/10.1038/s41558-021-01238-9.

Schaller, Nathalie, Alison L. Kay, Rob Lamb, Neil R. Massey, Geert Jan van Oldenborgh, Friederike E. L. Otto, Sarah N. Sparrow, Robert Vautard, Pascal Yiou, Ian Ashpole, Andy Bowery, Susan M. Crooks, Karsten Haustein, Chris Huntingford, William J. Ingram, Richard G. Jones, Tim Legg, Jonathan Miller, Jessica Skeggs, David Wallom, Antje Weisheimer, Simon Wilson, Peter A. Stott, and Myles R. Allen. 2016. 'Human Influence on Climate in the 2014 Southern England Winter Floods and Their Impacts'. *Nature Climate Change* 6 (6): 627–34. https://doi.org/10.1038/nclimate2927.

Schiemann, Reinhard, Panos Athanasiadis, David Barriopedro, Francisco Doblas-Reyes, Katja Lohmann, Malcolm J. Roberts, Dmitry V. Sein, Christopher D. Roberts, Laurent Terray, and Pier Luigi Vidale. 2020. 'Northern Hemisphere Blocking Simulation in Current Climate Models: Evaluating Progress from the Climate Model Intercomparison Project Phase 5 to 6 and Sensitivity to Resolution'. *Weather and Climate Dynamics* 1 (1): 277–92. https://doi.org/10.5194/wcd-1-277-2020.

Shepherd, Theodore G. 2014. 'Atmospheric Circulation as a Source of Uncertainty in Climate Change Projections'. *Nature Geoscience* 7 (10): 703–8.

———. 2016. 'A Common Framework for Approaches to Extreme Event Attribution'. *Current Climate Change Reports* 2 (1): 28–38. https://doi.org/10.1007/s40641-016-0033-y.

Shepherd, Theodore G., Emily Boyd, Raphael A. Calel, Sandra C. Chapman, Suraje Dessai, Ioana M. Dima-West, Hayley J. Fowler, Rachel James, Douglas Maraun, Olivia Martius, Catherine A. Senior, Adam H. Sobel, David A. Stainforth, Simon F. B. Tett, Kevin E. Trenberth, Bart J. J. M. van den Hurk, Nicholas W. Watkins, Robert L. Wilby, and Dimitri A. Zenghelis. 2018. 'Storylines: An Alternative Approach to Representing Uncertainty in Physical Aspects of Climate Change'. *Climatic Change* 151 (3–4): 555–71.

Sillmann, Jana, Theodore G. Shepherd, Bart van den Hurk, Wilco Hazeleger, Olivia Martius, Julia Slingo, and Jakob Zscheischler. 2021. 'Event-Based Storylines to Address Climate Risk'. *Earth's Future* 9 (2): e2020EF001783. https://doi.org/10.1029/2020EF001783.

Stott, Peter, Dáithí Stone, and Myles R. Allen. 2004. 'Human Contribution to the European Heatwave of 2003'. *Nature* 432 (7017): 608–10.

Stott, Peter, A, Nikalaos Christidis, Myles Allen, Randall Dole, Martin Hoerling, Chris Huntingford, Pardeep Pall, Judith Perlwitz, and Dáithí Stone. 2013. 'Attribution of Weather and Climate-Related Events'. In *Climate Science for Serving Society: Research, Modelling and Prediction Priorities*, edited by Ghassem R. Asrar and James W. Hurrell. New York; London: Springer.

Stott, Peter A., Nikolaos Christidis, Friederike E. L. Otto, Ying Sun, Jean-Paul Vanderlinden, Geert Jan van Oldenborgh, Robert Vautard, Hans von Storch, Peter Walton, Pascal Yiou, and Francis W. Zwiers. 2016. 'Attribution of Extreme Weather and Climate-Related Events'. *Wiley Interdisciplinary Reviews: Climate Change* 7 (1): 23–41.

Stott, Peter A., David J. Karoly, and Francis W. Zwiers. 2017. 'Is the Choice of Statistical Paradigm Critical in Extreme Event Attribution Studies'? *Climatic Change* 144 (2): 143–50. https://doi.org/10.1007/s10584-017-2049-2.

Stuart-Smith, Rupert F., Friederike E. L. Otto, Aisha I. Saad, Gaia Lisi, Petra Minnerop, Kristian Cedervall Lauta, Kristin van Zwieten, and Thom Wetzer. 2021 June 1–5. 'Filling the Evidentiary Gap in Climate Litigation'. *Nature Climate Change* 11, 651–55. https://doi.org/10.1038/s41558-021-01086-7.

Takayabu, Izuru, Kenshi Hibino, Hidetaka Sasaki, Hideo Shiogama, Nobuhito Mori, Yoko Shibutani, and Tetsuya Takemi. 2015. 'Climate Change Effects on the Worst-Case Storm Surge: A Case Study of Typhoon Haiyan'. *Environmental Research Letters* 10 (6): 064011. https://doi.org/10.1088/1748-9326/10/6/064011.

Tebaldi, Claudia, and Knutti Reto. 2007. The Use of the Multi-Model Ensemble in Probabilistic Climate Projections. *Philosophical Transactions of the Royal Society A* 365: 2053–75

Thompson, Allen, and Friederike E. L. Otto. 2015. 'Ethical and Normative Implications of Weather Event Attribution for Policy Discussions Concerning Loss and Damage'. *Climatic Change* 133 (3): 439–51.

Tradowsky, J. S., S. Y. Philip, F. Kreienkamp, Sarah F. Kew, Philip Lorenz, Julie Arrighi, Thomas Bettmann, Steven Caluwaerts, Steven C. Chan, Lesley De Cruz, Hylke de Vries, Norbert Demuth, Andrew Ferrone, Erich M. Fischer, Hayley J. Fowler, Klaus Goergen, Dorothy Heinrich, Yvonne Henrichs, Frank Kaspar, Geert Lenderink, Enno Nilson, Friederike E. L. Otto, Francesco Ragone, Sonia I. Seneviratne, Roop K. Singh, Amalie Skålevåg, Piet Termonia, Lisa Thalheimer, Maarten van Aalst, Joris Van den Bergh, Hans

Van de Vyver, Stéphane Vannitsem, Geert Jan van Oldenborgh, Bert Van Schaeybroeck, Robert Vautard, Demi Vonk and Niko Wanders. 2023. 'Attribution of the Heavy Rainfall Events Leading to Severe Flooding in Western Europe during July 2021'. *Climatic Change* 176, 90.

Trenberth, Kevin E. 2011. 'Communicating Climate Science and Thoughts on Climategate'. *Joint Presidential Session on Communicating Climate Change* 1–6.

Trenberth, Kevin E., John T. Fasullo, and Theodore G. Shepherd. 2015. 'Attribution of Climate Extreme Events'. *Nature Climate Change* 5 (8): 725–30.

Weisheimer, Antje, T. N. Palmer, and F. J. Doblas-Reyes. 2011. 'Assessment of Representations of Model Uncertainty in Monthly and Seasonal Forecast Ensembles'. *Geophysical Research Letters* 38 (16). https://doi.org/10.1029/2011GL048123.

Winsberg, Eric, Naomi Oreskes, and Elisabeth Lloyd. 2020. 'Severe Weather Event Attribution: Why Values Won't Go Away'. *Studies in History and Philosophy of Science Part A* 84 (December): 142–49.

WMO. 2019. *Statement on the State of the Global Climate in* 2018. Geneva: World Meteorological Organization.

Woollings, Tim. 2010. 'Dynamical Influences on European Climate: An Uncertain Future'. *Philosophical Transactions of the Royal Society A: Mathematical, Physical and Engineering Sciences* 368 (1924): 3733–56.

WWA. World Weather Attribution. 2018. *Heatwave in Northern Europe, Summer* 2018 https://www.worldweatherattribution.org/attribution-of-the-2018-heat-in-northern-europe/.

———. 2021. 'Pathways and Pitfalls in Extreme Event Attribution'. *World Weather Attribution* (blog). https://www.worldweatherattribution.org/pathways-and-pitfalls-in-extreme-event-attribution/.

———. 2022. 'High Temperatures Exacerbated by Climate Change Made 2022 Northern Hemisphere Droughts More Likely'. *World Weather Attribution* (blog) https://www.worldweatherattribution.org/high-temperatures-exacerbated-by-climate-change-made-2022-northern-hemisphere-droughts-more-likely/

Zappa, Giuseppe, Paulo Ceppi, and Theodore G. Shepherd. 2021. 'Eurasian Cooling in Response to Arctic Sea-Ice Loss Is Not Proved by Maximum Covariance Analysis'. *Nature Climate Change* 11 (2): 106–8. https://doi.org/10.1038/s41558-020-00982-8.

Zappa, Giuseppe, and Theodore G. Shepherd. 2017. 'Storylines of Atmospheric Circulation Change for European Regional Climate Impact Assessment'. *Journal of Climate* 30 (16): 6561–77.

Zscheischler, Jakob, Olivia Martius, Seth Westra, Emanuele Bevacqua, Colin Raymond, Radley M. Horton, Bart van den Hurk, AghaKouchak, Amir, Jézéquel, Aglaé, Mahecha, Miguel D., Maraun, Douglas, Ramos, Alexandre M., Ridder, Nina N., Thiery, Wim, and Vignotto, Edoardo. 2020. 'A Typology of Compound Weather and Climate Events'. *Nature Reviews Earth & Environment* 1 (7): 333–47. https://doi.org/10.1038/s43017-020-0060-z.

6 Toward a rectificatory policy mechanism for loss and damage

Rectificatory justice aims at repairing some negative situation by providing some normative force its historical origins. The intuition that there is something like rectificatory *climate* justice has appeared both in historical claims made by poor and developing countries, those most affected by climate change, and, sometimes, by political leaders from the Global North, who have posed some normative relevance to their historical role in causing climate change. Rectificatory climate justice for loss and damage has two aims. First, it aims at repairing the negative effects of climate change on people's capabilities and human rights. Second, it aims at distributing the burdens of reparation among those most closely connected to climate change–inducing activities, either because of their pollution (through the PPP) or because the benefits they acquired from pollution (through the BPP).

In this book, I have made two important contributions for an account of rectificatory climate justice. First, in Chapter 5, I presented different scientific methodologies that address the Causation Objection and argued that there are good reasons to reject basing the choice between those methods on whether any of them inherently overstates the effects of ACC. Yet the question remains of how to choose between those attribution methods to identify the effects of climate change. Similarly, in Chapter 4, I offered an argument based on the Continuity Account for why polluters should be made liable for loss and damage. However, I have not yet addressed the question of how liability could be distributed among polluters. Thus, two challenges remain: which attribution method ought to be preferred and how liability could be distributed among polluters within a policy mechanism for loss and damage. This chapter is dedicated to fill these gaps.

This chapter is structured as follows. First, I introduce the adequacy-for-purpose view as presented in the literature on philosophy of climate science and show why such a view could be applied to attribution methods. Second, I provide an adequacy-for-purpose argument for choosing between attribution methods for rectificatory climate justice purposes. Although this argument is non-conclusive, I take it to provide some relevant reasons to prefer one approach, namely, the probabilistic approach, over the alternative storyline approach. Second, I explain how a policy mechanism for loss and damage could function following the input provided by attribution studies and how it would circumvent some objections raised against rectificatory justice approaches. Finally, I address the problem of political

DOI: 10.4324/9781003399889-6

feasibility of a historical responsibility-based account and develop a somewhat speculative argument to counter this objection. With these insights, my main aim is to contribute to the development of a policy mechanism for rectificatory climate justice for loss and damage.

6.1 Toward an adequacy-for-purpose view for attribution methods

Science is a purpose-oriented activity. A growing number of philosophers of science agree that the purpose of science is not merely representational. That is, the purpose is not only to develop a certain imagine of the world but rather to contribute to the achievement of certain general purposes. This idea has also influenced the development of the philosophy of climate science.

The concept of adequacy-for-purpose emerges as an alternative to the concepts of verification and confirmation, which are associated with the understanding of science as a representational tool, for dealing with model evaluation. The concept of confirmation emerged in opposition to the use of the concept of verification in the context of climate models. As Oreskes, Shrader-Frechette, and Belitz (1994) argue, verification implies the truth of the proposition one aims to demonstrate. But verification is only possible in closed systems such as mathematical models, not in open systems containing empirical hypotheses, such as representations of the natural system.[1] Models are never closed because they require changes in boundary conditions and auxiliary hypotheses (e.g., additional assumptions, interferences, and input parameters) used to represent the natural system. All scientific inferences about climate involve a combination of models and observations that depend on empirical data. For this reason, the system is always open and cannot be verified.[2] This led Oreskes, Shrader-Frechette, and Belitz to propose the concept of confirmation as an alternative approach to discussing the evaluation of climate models. Instead of truth, confirmation takes empirical adequacy as the goal for climate models.

Proponents of the adequacy-for-purpose view have argued that

> even if a model is viewed as a complex hypothesis about the workings of a target system. . ., it is usually misguided to seek to confirm (or disconfirm or falsify) that hypothesis since it is usually known from the outset to be false; some of the model's assumptions are known to be highly idealized or simplified, to appeal to fictional entities, and so on.
>
> (Parker 2020, 458)

The concept of adequacy-for-purposes goes one step further in detaching climate models from their exclusive consideration as representational tools (Parker 2009, 2020).

The concepts of both verification and confirmation take the representation of the empirical world as the exclusive (or at least primary) role of climate models. An adequacy-for-purpose view takes climate models as tools that can be valuable for purposes other than representation. The idea is that 'model evaluation [under an adequacy-for-purpose-view] seeks to learn whether a model is adequate or fit for particular purposes' (Parker 2020, 458). This means that model evaluation needs to consider not only whether a particular model represents reality in an accurate way but also (and independently) whether it is adequate for the circumstances in which it is intended to be used. That includes the target, the user, the methodology, the goal, the research questions, etc.

Parker (2020, 462) considers that the most extended understanding of adequacy-for-purpose is associated with reliability in a type of use. She defines this as follows:

> A tool M is adequate-for-p if and only if, in C-type instances of the use of M, purpose P is very likely to be achieved.

Here, whether M is adequate-for-p depends upon the context (C) in which the tool is intended to be used. C-type instances include things like the characteristics and abilities of the users (U), methodology (W), background contexts (B), or the required degree of fidelity (T) (Parker 2020, 464). Parker provides an example of how her adequacy-for-purpose view could work for the evaluation of climate models. She asks us to consider a computer model M designed to simulate the local rise in sea level that occurs along the US coast when a hurricane is nearby. The model developers would like their model to predict whether the rise in sea level at specific coastal locations will exceed the hazard parameters. The context of interest involves forecasters U at the US National Hurricane Center, who follow a methodology W for their predictions and within a certain background B (for instance, their computers operate normally without power cuts). A model M would be adequate for predicting the rise in sea level at that particular US coast if, whenever the forecasters use M in a way W and within B, they are likely to predict correctly whether the sea level will exceed the hazardous thresholds.

A second concept central to the adequacy-for-purpose approach is fitness-for-purpose. The concept of fitness-for-purpose serves to rank multiple models that are all adequate for a certain purpose. To that end, the concept of fitness-for-purpose, unlike that of adequacy-for-purpose, is scalar; that is, it admits degrees. If we assume that a certain purpose has a complex structure with a rank order, where Pmin is equivalent to being minimally and Pmax equivalent to being maximally adequate for the purpose, then we can assess the relative fitness of a model. Parker (2020, 463) defines fitness-for-purpose as follows:

> A model M's FITNESS-FOR-P is greater to the extent that M is ADEQUATE-FOR-P for a higher-ranking member of P = {Pmin, . . ., Pmax}.

Consider her example of Illinois high school students' knowledge of world geography and a geography teaching model. For increasing students' world geography knowledge, even a small increase in knowledge counts as a success, but, of course, larger increases are more desirable. It might then be the case that a teaching model M1 increases the high school students' knowledge of world geography more in most instances than a different model M2. M1 thus has a greater fitness-for-purpose than M2. In a nutshell, the concept of fitness-for-purpose allows for a comparison between 'tools', all of which are adequate for a certain purpose, to assess which one is better suited for that purpose. That is, the concept of fitness-for-purpose allows us to rank 'tools' according to the level of their adequacy for a certain purpose in a certain context.

This brief description of adequacy-for-purpose and fitness-for purpose should suffice to indicate how these concepts can be applied to the evaluation of attribution methodologies to seek reparations for losses and damages. Though both concepts are involved in my discussion, I refer to the overall approach as the adequacy-for-purpose view, with the understanding that evaluating adequacy-for-purpose may also involve evaluating fitness-for-purpose.

6.2 An adequacy-for-purpose view for rectifying climate injustice

Two points should be emphasized from the previous discussion. First, the concept of adequacy-for-purpose can be applied to the evaluation of any tool, whether a hammer, a computer, a climate model, or an attribution methodology. In other words, there is nothing inherent in the adequacy-for-purpose view that restricts this approach only to models. We could apply this framework to any purpose-directed scientific activity (Lusk and Elliott 2022). Hence, the shift from adequacy-for-purpose in model evaluation to adequacy-for-purpose in evaluating scientific methodologies for attribution purposes is justified. With this reminder, we can move on to analyze whether, to what extent, and under which conditions and circumstances, the two scientific methodologies under discussion – the probabilistic approach and the storyline approach – can be adequate for attributing EWE to anthropogenic forcing.

Second, the adequacy of these scientific methodologies for attributing EWE to ACC varies depending on the context of use. Here, recall that the instance of use under consideration is liability for loss and damage. But whether attribution studies can be adequate for the purpose of attributing EWE to anthropogenic climate change in the context of reparations for loss and damage will also depend upon the normative system under consideration. Thus, the conditions of adequacy-for-purpose of attribution studies should be qualified depending on their specific instance of use.

We can consider two domains within which attribution studies can serve the purpose of attributing EWE to anthropogenic forcing in order to assign liability for the costs of losses and damages caused by climate change: the legal domain and the policy domain. Each of these domains has different adequacy-for-purpose conditions. From a legal perspective, whether attribution studies are adequate for the purpose of *legally* attributing EWE to anthropogenic forcing depends on whether this purpose is likely to be achieved within the constraints imposed by our legal systems. In turn, this will depend upon the different legal systems' requirements for proving causation. If attribution studies cannot provide the kind of answers required to prove causation in our legal systems, then the answer is negative. If they can, then the answer would be positive. Moreover, it might be the case that different scientific methodologies are better suited than others to meet the empirical legal demands. Whether there is a plausible interpretation of the legal texts that may accommodate the statements provided by any methodological approach to attribution studies is a matter that is open for legal debate. In any case, this question lies beyond the scope of this book.

In the policy domain, the question of adequacy-for-purpose looks slightly different. In fact, in this context, we may turn the adequacy-for-purpose question around. That is, in the legal domain, we compare the empirical demands of our normative system (i.e., a legal system) with the information provided by attribution studies to see whether the results of attribution studies fit those demands. However, in the policy domain, our purpose is not to fit the demands of a given normative system (such as the legal system) but rather to *develop* a normative system that serves certain (moral and justice-related) purposes. However, the development of a new normative system must take into consideration *the best empirical information* available for the purpose at hand, thereby establishing new empirical standards for meeting the purposes we want to achieve.

Recall that the purpose of a climate policy mechanism for loss and damage is twofold. First, such a mechanism should be able to produce *the best approximation* of loss and damage caused by climate change inducing activities (i.e., it should be able to provide the closest approximation to climate change–related harm). Second, this policy mechanism should also serve the purpose of rectifying climate injustice, which requires repairing an injustice by accounting for the sources of the injustice, namely, emission-generating activities. This book engages with one approach to rectifying climate injustice in this sense, which is to make polluters liable for loss and damage via the application of a principle of historical responsibility. In other words, the purpose at stake here is to achieve the best *identification of climate harm* for which *liability* will then be distributed. Now, the relevant question is whether the existing attribution methods are adequate for this purpose and, if so, which one is fitter for that purpose.

We have good reasons to believe that both the probabilistic event attribution and the storyline approach are adequate for the purpose of identifying climate harm for which liability might then be distributed and thus are adequate for informing a rectificatory justice policy mechanism for loss and damage. Both the probabilistic approach and the storyline approach are widely accepted scientific methods. Recall that proponents of the storyline approach have always considered their preferred methodology to be compatible with the probabilistic approach. At the same time, despite the initial hostile reception of the storyline approach among the PEA community, this alternative approach has become widely accepted among the climate science community, as is reflected in the last IPCC report (IPCC 2021). For this reason, both approaches can be considered the best tools we currently have to achieve the best approximation of loss and damage caused by climate change inducing activities. At the same time, both approaches could in principle be used to distribute liability for loss and damage, since both approaches link GHG emissions with certain climate change–related harm. Thus, both approaches could be used to distribute liability for climate harm among those who contribute to climate change. In principle, therefore, both accounts seem to be adequate for the purpose of informing a rectificatory justice policy mechanism for loss and damage.

The question remains as to which attribution method is fitter for that purpose. Providing a complete answer to this question would be necessary in the few years to come, as attribution studies develop, and their social and political purpose becomes more and more relevant. However, such a task would require much more than I can do in this chapter and in this book. Nevertheless, I believe there are some reasons speaking in favor of fitness-for-purpose of the probabilistic approach in the context of rectificatory climate justice. Here, I would like to present these reasons, with the caveat that by no means I take them to solve the problem at hand. Further research would need to be done to conclude what approach is fitness-for-this purpose all things considered.

Let me start by revisiting the difference between the two methods in terms of their research questions. Recall that the probabilistic approach takes a certain event as a token of a class of EWEs and asks the following research question: How much did ACC increase the probability or risk of a specific type of event? The answer to this question would be of this sort: 'ACC has increased (or decreased) the probability of occurrence of this type of EWE by a factor X (probability ratio)'. In contrast, the storyline approach takes certain dynamic variables as fixed and asks about changes in magnitude. For this approach, the relevant research question is: '*Given* certain dynamic variables, how much has ACC increased the magnitude of this particular event?' Accordingly, the answers attached to this methodology are of this sort: 'ACC increased the EWE's magnitude by a value or factor of x, *assuming* that the dynamic variables are so and so'. In the following, I will be arguing that the nature of the questions asked by the probabilistic approach and the corresponding answers make it fitter for the purpose of liability than those involved in the storyline approach because they resemble our liability thinking as reflected in legal practice.

The use of probabilistic and risk assessment for attributing certain effects to certain causes is not entirely new. Legal practice already included this possibility in analogous contexts to climate change. An example of this is the use of probabilistic causation and risk assessment in liability concerning the effects of exposure to asbestos on people's health. For instance, in the legal case *Heneghan v Manchester Dry Docks Ltd*, the company was declared responsible for its employee's lung cancer due to his exposure to asbestos in the workplace.[3] The evidence taken as relevant for establishing this causal connection was framed in terms of the increase in the risk of developing cancer due to exposure to asbestos. More specifically, the company was held liable for causing cancer because the exposure to asbestos had doubled the risk of developing cancer. This led the court to conclude that, on the balance of probabilities, the employee's exposure to asbestos was the cause of his lung cancer. The underlying reasoning here involves two assumptions. First, the worker must be compensated for suffering the impacts of cancer, and second, compensation must be provided by those who have likely caused it. A risk assessment showing that his exposure to asbestos has doubled the risk of developing cancer indicates that we are more likely to be right than wrong in attributing his cancer to his exposure to asbestos. Arguably, this is taken as a sufficient reason to make the company liable for the negative effects of his exposure to asbestos.[4]

Cases such as *Heneghan v Manchester Dry Docks Ltd* show that our understanding and practice of attributing liability and causation is already framed in the terms proposed by the probabilistic approach.[5] In a sense, probabilistic assessments of this sort are already present in our social understanding of what it means to attribute an effect to a cause in contexts of liability for damages. In my view, it is this similarity between the probabilistic approach and existing causation claims in certain liability contexts that makes this approach fit for attributing EWEs to human actions in the context of liability for loss and damage. The fact that the probabilistic approach can provide the kind of assessments accepted in other legal cases makes it fit for this context.[6]

One might object that my argument seems to be contradictory. Previously, I established a distinction between the legal and the policy domain to argue that my focus remains within the policy domain. But now I turned to legal sources to make a case in favor of the probabilistic approach. However, these two claims are not incompatible. I argued earlier that, to develop a policy mechanism, one need not adhere to the established empirical standards within law. Instead, the demandingness of the empirical standards of a policy mechanism can be shaped depending on its role and the function that it is intended to serve. However, this does not mean that the legal domain cannot provide some guidance on the kind of methods that should be preferable in certain contexts. In my view, the use of the probabilistic approach in certain legal contexts associated with liability for damages is a reason to favor the probabilistic approach over its alternative for its 'fitness-for-liability'. Of course, further research is needed to determine whether these advantages conclusively show that we ought to prefer the probabilistic approach over the storyline approach all things considered.

Proponents of the storyline approach might press a second objection. They might argue that the forensic and deterministic reasoning present in the storyline approach also reflects forms of causation present in liability thinking and legal practices. Thus, the same reasoning could be applied to support the fitness-for-liability of the storyline approach, or even stronger, to argue for the superiority of the fitness-for-liability of the storyline approach, given that its reasoning resembles stronger forms of causation. For instance, Lloyd and Shepherd have argued that the storyline approach can establish deterministic 'but for' causation, which is a stronger form of causation recognized in legal cases than the kind of probabilistic causation involved in toxic tort law cases (Lloyd and Shepherd 2021, 28). The storyline approach provides evidentiary narrative, in a forensic sense, about the role of each factor in a single impact, potentially including other factors apart from merely physical ones (e.g., the so-called confounding factors such as vulnerability and exposure). Lloyd and Shepherd exemplify this kind of narrative with a figure similar to this one:

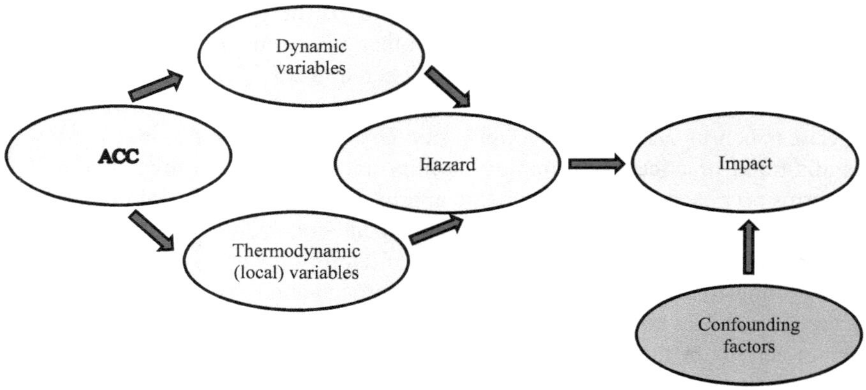

Figure 6.1 Storyline evidentiary narrative as represented by Lloyd and Shepherd. Arrows represent casual influence.

Source: Lloyd and Shepherd (2021)

Take a liability case involving a car accident and damages suffered by a pedestrian. Here, when we assess whether the damages suffered by a pedestrian were caused by a hitting car, we assess whether these damages would have happened if it were not for the car. In other words, we assess whether the damages happened 'but for the car'. We do not assess whether the probability of the pedestrian suffering damages increased by a certain degree due to the driving car. Similarly, the storyline approach builds 'but for' narratives such as: had it not been for ACC, the thermodynamic (local) variables would have been different and the magnitude of the event (described in terms of hazard or impact) would have been, for instance, lower. As we saw, such a result would be achieved by comparing the values of thermodynamic (local) variables with ACC and with natural variability only. The role of factors (such confounding factors) in the magnitude of the event can also be considered in this narrative. For instance, if the target population would have not had building of this sort, the damages caused by the changes in thermodynamic factors would have been so and so.

However, and importantly, in this narrative, the potential effects of ACC on dynamic variables are ignored. Instead, dynamic variables are fixed and only the effects of ACC on thermodynamic variables are considered. For this reason, Llyod and Shepherd claim that 'the storyline approach can establish deterministic "but for" causation, *albeit in a conditional manner*' (Lloyd and Shepherd 2021, 28). In other words, by fixing the dynamic variables, the storyline approach ignores how the changes in dynamic variables affected the event. Thus, the storyline approach works rather often as in Figure 6.2:

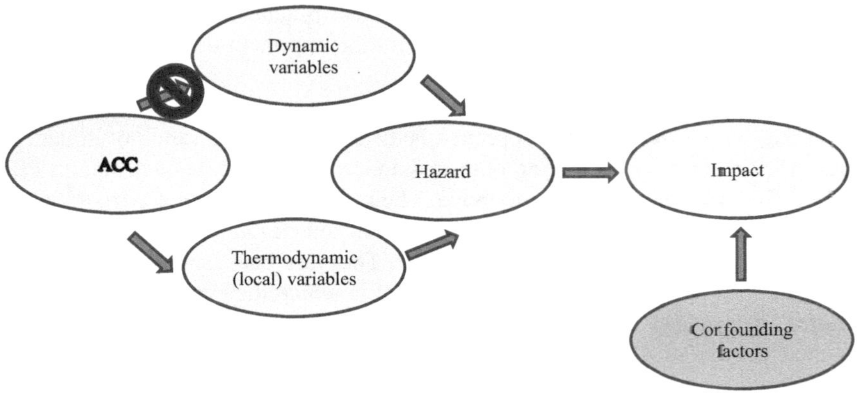

Figure 6.2 Storyline evidentiary narrative. Arrows represent casual influence.

Source: My own representation

Although this conditional manner does not, at first, seem to invalidate the claim that the storyline approach also reflects our standard ways of thinking about causation and liability, I will argue that this initial perception is misguided. Instead, I argue that the conditional characteristics of the storyline approach make this approach depart from standard liability reasoning and thus less fit for liability purposes.

The main reason for believing that the conditional nature of the storyline approach does not present any disadvantage is that, at the end of the day, any causal connection is conditioned on certain 'fixed' factors. When we compare the actual world with a counterfactual one, we always take some background factors to be 'fixed'. When we assess whether certain injuries a pedestrian caused by a car, we compare the actual world where the pedestrian was hit by the car with a world in which all factors except for the car hitting the pedestrian were present. We 'fix' things such as the pedestrian's age or the weather. If in the counterfactual world in which the car is absent, the pedestrian had not suffered those injuries, then we conclude that the injuries of the pedestrian were caused by the car. Similarly, in many attribution studies, including those carried out with the probabilistic approach, some *background factors* are taken as fixed. For example, the same volcanic forcings on climate are taken as fixed when comparing the probability of an EWE in a world with and without ACC. Similarly to the car example, this is done to ensure that differences between p1 and p0 reflect only the additional impact of human influence (Stott, Karoly, and Zwiers 2017, 145), with the exception that the causal connection established here is a probabilistic one. Recently, even some traditional proponents of the probabilistic event attribution have argued that attribution methods often differ in the *degree* of conditioning *rather than fundamentally* (Stott, Karoly, and Zwiers 2017; Otto 2023). For those reasons, neither the conditional nature of the storyline approach represents an exception to how attribution studies are carried out in comparison to other attribution methods, nor does it differ from how liability assessments are carried out in legal practice. Hence, the argument would go: the conditional nature of the storyline approach does not constitute a disadvantage for its fitness-for-liability.

However, this counterargument is somewhat misguided. For the conditioning happening within the storyline approach is different in *nature* (and not in degree) from the conditioning happening both in standard liability cases and within PEA studies. Consider the car accident above. The kind of things that are 'fixed' when considering the role of the car in the pedestrian injuries are *background factors*, such as the pedestrian's age or the weather. They are background factors in the sense that they define the kind of environment in which the event occurs and their features they do not depend on the cause of the event under consideration. For instance, the age of the pedestrian does not depend on the presence or absence of the car, nor on the driver's behavior. Similarly, PEA studies fix background conditions such as volcanic forcings, which bear no relationship of dependency with the cause of the events at stake, namely, ACC.

In contrast, the storyline approach fixes factors that are not part of the background conditions under which environmental impacts from climate change happen, not at least in the sense described earlier. Remember that the storyline approach fixes the dynamic variables of an EWE. However, we know that these dynamic variables are affected, or at least are very likely to be affected, by ACC and thus depend on ACC. Whereas the pedestrian's age is not dependent on the driver's behavior, dynamic variables are dependent on ACC. Likewise, whereas volcanic forcings are not dependent on ACC, dynamic variables *are* dependent on ACC. For this reason, dynamic variables are not background factors of the event, or at least not in the same way as the pedestrian's age or volcanic forcings are in standard liability cases and in PEA studies respectively. Further, by fixing the dynamic variables, the storyline approach ignores some of the direct effects (changes in dynamic variables) of the cause of the event (ACC), whereas standard liability cases do not ignore the direct effects of the cause of the event (faulty driving) by fixing the age of the pedestrian in the example above. Neither does the PEA ignore the direct effects of the cause of the event (i.e., ACC) by fixing volcanic forcings.

These remarks cast doubt on the two claims grounding the possible counterargument above. First, the kind of 'conditional' but for narrative involved in the storyline approach differs fundamentally from standard 'but for' narratives present in standard liability cases. Second, the storyline approach differs fundamentally in the kind of conditioning employed (and not as a matter of degree) from the probabilistic approach.[7] These two conclusions give us reasons to consider that, at least in principle, the type of causation evoked by the probabilistic approach resembles more closely the type of liability thinking present in the law than the storyline approach. This would make the probabilistic approach more fit-for-liability than the storyline approach.

However, three caveats are in order before closing this section. First, note that the conditional results of the storyline approach can be combined with an analysis of how changes in circulation could affect the estimation about the effects of ACC (Lloyd and Shepherd 2021, 27–28). Second, recall that, as I argued in Chapter 5, the probabilistic account often uses proxy-definitions that tend to leave aside or downplays dynamic factors, thereby operating in practice in similar ways to the storyline approach. Third, recall that the problems concerning the representative power of climate models for modeling dynamic changes might make the probabilistic approach not even adequate for the purpose of rectificatory justice. Indeed, in those instances, the storyline approach might be better able to represent the best approximation to climate change–related harm.

These three remarks contribute to emphasize two important conclusions of the discussions I offered in this section. First, the probabilistic approach might be fitter for liability purposes at an abstract level given the kind of causation it evokes and its relationship with legal thinking. However, this does not mean that this approach would be fitter-for-liability in all cases. Second, relatedly, the argument offered in this section remains highly inconclusive. Even though the probabilistic approach might be fitter-for-liability at an abstract level, this does not mean that this approach ought to be preferred, all things considered, to the storyline approach for liability purposes. Although answering such a question lies beyond the scope of this book, these remarks are nevertheless important because any attempt to provide a conclusive answer will need to take them into consideration.

6.3 Distributing liability and achieving rectificatory justice

We now know what kind of considerations are relevant for choosing among attribution methods to identify climate change–related harm and thus loss and damage. However, two issues must be clarified in order to explain how a policy mechanism of this sort could achieve rectificatory justice for loss and damage. First, we need to know how to distribute liability for climate change–related harm. Second, we need to explain how the identification of harm and the distribution of liabilities can achieve rectificatory climate justice given uncertainties surrounding climate change scientific information.

Let us start with the first issue. Some basic facts about climate change may appear to challenge the possibility of attributing liability for loss and damage. As is well known, even if we can attribute certain instances of environmental harm to climate change, there is no discernible link between the emissions of a particular state and specific instances of loss and damage. Greenhouse gases are mixed in the atmosphere and become evenly distributed shortly after being emitted, irrespective of their geographical location. Moreover, many of those gases are 'stock pollutants' in that the changes they force in the climate system are a function of their accumulation over years. These two facts make it impossible to attribute certain damages to specific agents. That is, these two facts undermine the possibility of ascribing any particular causative event occurring in a certain geographical location specifically to emission-generating activities of a particular state (Page 2012).

To solve this problem, we can again appeal to the possibilities of legal practice to shape policymaking. The difficulties of finding a direct link between the contributions of individual actors to a general problem and the specific harms caused by that problem are not exclusive to climate change, and they have been dealt with in legal practice before. For instance, in the years between 1947 and 1971, millions of American women took an estrogen-type drug, DES, to prevent miscarriages. In 1971, DES became contraindicated after research found an unquestionable link between DES and cancerous and precancerous vaginal tract abnormalities in prenatally exposed daughters of DES users. Some years later, millions of women faced the possibility of developing vaginal and cervical cancer but were unable to identify which of several hundred companies manufactured the particular pills their mothers took. Despite these difficulties, the California Supreme Court concluded that manufacturers could be made liable for those harms using what became known as the market share mechanism. In a nutshell, the market share mechanism distributes liability for certain harms according to a company's proportional contribution to the entire production of the drug sold by all companies. Other courts have applied this method of attributing liability for harm caused by DES and other products. (Sheffet 1983; Priest 2010)

Determining whether or not a market share mechanism could contribute to distributing legal liability lies beyond the scope of this book and my field of expertise. However, this mechanism could be used to distribute liability within a policy mechanism for loss and damage. In this way, again, legal practice can inform policymaking in the context of loss and damage.

As I envision it, a policy mechanism for loss and damage could incorporate, on the one hand, knowledge gained from attribution studies to identify loss and damage. On the other hand, such a policy mechanism would incorporate a fund whereby money is collected proportionally according to each state's contribution to climate change (following the PPP, as I argued in Chapter 4) and distributed in the form of reparations to those suffering loss and damage as identified by attribution studies.

Of course, there are some uncertainties surrounding both attribution science and distribution of liabilities. On the attribution science side, there is admittedly always even a minimal possibility that climate change in fact did not cause the attributed loss and damage. For example, following the probabilistic approach, even if it is concluded that climate change has increased the probability of an event happening by 200%, there is still a small chance that climate change did not in fact cause the event. Following the storyline approach, we do not know the true changes in dynamic conditions due to climate change and thus we cannot be 100% certain that the increases in magnitude attributed to climate change are accurate. Similarly, we cannot know which agents' emissions specifically caused loss and damage at a certain moment in a particular geographical location. Hence, when we use a market share mechanism for distributing liability, the distribution of liability in particular instances of loss and damage most likely does not truly reflect the contribution of each state to that climate change–related harm. Some might think that this poses a significant problem for a rectificatory justice account, since we cannot be sure that rectificatory justice for loss and damage is achieved with this policy mechanism.

It might indeed be that particular attribution statements and particular distribution of liabilities are (partly) misguided. However, this does not undermine the achievement of rectificatory justice. Here, I follow Rawls in arguing that justice is a virtue of institutions, and thus the primary subject of justice is 'the way in which social institutions distribute fundamental rights and duties'. In other words, justice is achieved if, overall, social institutions realize a distribution of rights and duties according to the purposes they are intended to fulfill. A policy mechanism for loss and damage achieves rectificatory justice if it is designed in the best possible way to repair the harmful effects of climate change, thereby fulfilling reparatory rights, while providing some normative relevance to its historical origins insofar as this is possible, thereby distributing rectificatory duties to those closely connected to the source of climate injustice. As I have argued in this book, a policy mechanism for loss and damage based on the PPP achieves this aim. Its institutional design according to these two parameters enables the policy mechanism for loss and damage to achieve rectificatory justice, not its particular performance in every single instance of loss and damage.

This explanation should soften some concerns about rectificatory approaches being 'too time-consuming and costly' and 'difficult to be relied upon to produce just outcomes . . . given the enormous social and economic inequalities between the two parties [involved]' (Boran and Heath 2016, 248). These concerns derive from a tort-law interpretation of rectificatory justice where it is assumed that victims need to have the necessary resources to bring polluters to court to achieve their justice demands (Boran and Heath 2016, 245). However, as we have seen, this should not be a concern for the approach I have presented here. A policy mechanism enables climate justice to be achieved in a more stable, organized, and clear manner so that individuals have the security that resources will be available to them if they suffer losses and damages from climate change.

To close this section, I would like to introduce and answer two final objections to my account. The first objection concerns skepticism about the forward-looking potential of a rectificatory approach. For instance, some scholars argue that reparations are 'inefficient as a way forward against the effects of climate change' and that they cannot 'be effective in securing improvement and infrastructural resilience [and] fall short of providing politically viable guidance to multilateral cooperation on adaptation' (Boran and Heath 2016, 247). However, the concern that reparations cannot ensure adaptation seems to be beside the point. Adaptation and L&D are two different responses to climate change and thus can be kept distinct when it comes to their policy design. If adaptation is not ensured with reparations for loss and damage, we might simply need a different policy mechanism for adaptation. Moreover, an approach that keeps adaptation and L&D separate can still address these concerns by enabling synergies between adaptation and L&D. For instance, as some scientists have suggested, the empirical data gathered for L&D purposes might also be useful for adaptation planning (Stott, Karoly, and Zwiers 2017).

The second objection concerns political feasibility. Some philosophers and political theorists have argued that a rectificatory justice approach may generate a polarized atmosphere among political parties, which, in turn, would hinder a consensus in political negotiations on climate policy. In the absence of that consensus, we would not reach any meaningful solution to climate change or to the problems associated with loss and damage (Posner and Weisbach 2010; Moellendorf 2014;

Boran and Heath 2016; Wallimann-Helmer et al. 2018). Moreover, this concern has gained political momentum in recent years because of paragraph 52 of decision 1/CP.21 of the Paris Agreement, which rules out any basis for liability and compensatory claims. This paragraph has been interpreted as a rejection of rectificatory approaches to climate justice of the sort I have defended in this book. This is what I call the Political Feasibility Objection.

Before I engage with this problem in more depth in the next section, some preliminary remarks will be useful here. First, invoking rectificatory justice principles might have made agreements more difficult thus far, but negotiations also depend on the political actors at any given time. It is not written in stone that future political actors will not have other incentives and motivations. There might be reasons to believe that rectificatory justice arguments will be more successful in the future. Second, even if mechanisms based on historical responsibility do not generate the necessary consensus in negotiations because some parties reject them, the same holds true for mechanisms that ignore historical responsibility. The historical roots of climate change trigger resentment among those who suffer their consequences. Any policy mechanism that does not acknowledge the role of emission-generating activities in the harm associated with loss and damage is likely to be rejected by countries vulnerable to climate change. Therefore, rejecting historical responsibility is not likely to facilitate an agreement.

In the next section, I elaborate a more detailed (and somewhat speculative) response to this objection.

6.4 The political feasibility objection

Feasibility issues are important if we aim to develop arguments with impacts in real-world policy, which should be a desideratum of political philosophy (Rawls 1999, 2001; Brennan and Pettit 2007; Kenehan 2017). This desideratum is contained in the Rawlsian idea of developing theories that are realistically utopian. The idea is that our philosophical theories should have real possibilities of success ('realistically') but should at the same time be morally desirable ('utopian'). The idea of political feasibility touches upon the first of those elements. In climate justice, the problem of political feasibility is highly relevant if we aim to develop theories that have real-world impacts on climate policies.

My aim in this section is to spell out the grounds of the Political Feasibility Objection and to provide a somewhat speculative argument to challenge this objection. In a nutshell, I argue that some empirical evidence might speak against the motivational concerns grounding the Political Feasibility Objection. Specifically, I argue that rectificatory arguments might be more likely to increase support for policies aimed at repairing loss and damage and that this might have an indirect influence on the political motivation of the relevant political actors to provide reparations for loss and damage based on historical responsibility. At the very least, these findings provide a reason not to reject rectificatory approaches outright on feasibility grounds.

6.4.1 Introducing and exploring the feasibility concern

The Political Feasibility Objection claims that a rectificatory justice approach may generate a polarized atmosphere among political parties that will hinder political consensus and a solution to climate change and/or to loss and damage. Thus, a rectificatory justice approach is unfeasible.

Let us start with a detailed examination of the claim that 'support for L&D based on compensatory justice may be currently politically unfeasible' (Wallimann-Helmer et al. 2018). Pablo Gilabert and Holly Lawford-Smith (Gilabert and Lawford-Smith 2012, 5) begin their analysis of feasibility by spelling out its different components: 'It is feasible for X to φ to bring about O in Z. In this formulation, X ranges over the relevant agents; φ refers to the relevant action; O refers to the outcome and Z refers to the contextual factors of space and time'.

According to this formula, the elements of the unfeasibility claim in the case of loss and damage could be interpreted as follows:

- X refers to the COP within the UNFCCC;[8]
- φ refers to the acceptance and use of rectificatory arguments based on historical responsibility;
- O refers to repairing loss and damage from climate change now and in the future;
- and Z refers to the current and mid-term international order.

Therefore, the objection's claim is that it is unfeasible for the UNFCCC to accept and use rectificatory arguments based on historical responsibility to repair loss and damage from climate change in the context of the current and mid-term international order.

Before going into further details, we should also note a distinction between two forms of unfeasibility. First, something might be unfeasible because it meets 'hard constraints'. Hard constraints are constraints that 'will always be constraints' (Gilabert and Lawford-Smith 2012). These constraints rule out possible actions. Physical or biological constraints are good examples of hard constraints. Gravity and my features as a human being are a constraint on my ability to fly to the moon on my own power. Only major changes in human enhancement technologies, which are virtually impossible to achieve in my lifetime, could change that fact. Therefore, gravity will always be a hard constraint on my ability to fly to the moon. This is why flying to the moon is unfeasible for me in my lifetime.

Soft constraints are constraints that create difficulties for certain options. They do not rule them out but make them comparatively less feasible. Examples of these constraints are economic, institutional, cultural, motivational, or psychological constraints. Unlike hard constraints, soft constraints involve a probabilistic component. When we refer to soft constraints, we do not mean that the proposals affected by them are impossible to carry out. Rather, we think that 'the probability of their bringing it about is not high enough' (Gilabert and Lawford-Smith 2012, 6). Soft constraints are also malleable: something can be done to overcome them. Whereas hard constraints rule out possible actions, soft constraints indicate the difficulties of acting.

Hard and soft constraints render proposals infeasible in different ways. Hard constraints make certain options infeasible in a binary sense, and soft constraints make certain options infeasible in a scalar sense. According to Gilbert and Lawford-Smith, 'a proposal is feasible in the binary sense only if it does not violate hard constraints, and is more feasible in the scalar sense the more it accommodates soft constraints' (Gilabert and Lawford-Smith 2012, 7). More specifically, they define binary and scalar feasibility constraints as follows:

Test 1: Binary: It is feasible for X to φ to bring about O in Z if X's φ-ing to bring about O in Z is not incompatible with any hard constraint.

Test 2: Scalar: It is more feasible for X to bring about O_1 than for Y to bring about O_2 when it is more probable, given soft constraints, for X to bring about O_1 given that he or she tries, than it is for Y to bring about O_2 given that he or she tries.[9]

With these distinctions in mind, we can now explore the kind of constraints that give rise to Political Feasibility Objection. The first important thing to note is the nature of the relevant agent involved, in this case the COP is the relevant agent (similar to Kenehan 2017). This is so because developing an international policy instrument for loss and damage requires an international agreement within the UNFCCC, and COP is the supreme decision-making body of the UNFCCC where all the States that are Parties to the Convention are represented and where decisions about climate policies are undertaken. Note also that the COP is not a unified agent. Instead, it comprises different parties who need to discuss and agree on a specific matter for any decision taken by the COP.

The nature of the COP already reveals one of the constraints that give rise to the feasibility concern, namely, an institutional constraint. The institutional constraint arises from two factors. First, for the COP to make a decision such as the one involving the rectificatory grounds of loss and damage policies, there needs to be an agreement between the Parties. However, second, the relevant Parties have different and conflicting interests and motivations. To simplify matters, we can divide the Parties between developed and industrialized nations (Annex I countries) and less developed and vulnerable nations (non-Annex I countries). As Kenahan has observed, historically speaking, developed and industrialized nations (particularly, the United States) have rejected rectificatory arguments based on historical responsibility, while less-developed and vulnerable nations have pushed for an agreement that reflects historical responsibility (Kenehan 2017, 199–204). The UNFCCC has clearly reported this disagreement within the negotiations:

> For some [parties], historical responsibility was central to the discussions on a formulaic approach, and this issue has to be resolved in terms of responsibility for current impacts before discussing the responsibility of non-Annex

> I Parties. For others, it was not seen as an adequate measure of equity because it is complex, statis and includes a large number of variables.
>
> (UNFCCC 2012)

To summarize, the institutional setting represents a constraint on the COP's ability to develop a policy instrument to repair loss and damage based on a rectificatory framework. That constraint is mainly given by the necessity of reaching an agreement among parties with different interests and incentives. Therefore, any solution is blocked until such an agreement can be reached.

Once we know what the main constraints are, it is time to consider the sense in which they render a rectificatory approach infeasible. The constraints we face are soft because they concern institutional mechanisms and can potentially be changed. Therefore, a rectificatory approach would be more or less feasible in a scalar sense depending on how it accommodates the constraints at stake. But in order to work out the ways in which those constraints can be accommodated, we must delve a bit deeper into their nature.

I noted earlier that this institutional constraint emerges because of the need for an agreement as well as because of competing interests among the different parties. This is what hinders the development of policy instruments based on a rectificatory framework. However, note that these two factors would not pose a major problem if all the Parties involved would *try* to embrace a rectificatory framework. That is, there are no further institutional constraints that would hinder framing loss and damage under rectification beyond the fact that *some* Parties are not motivated to accept that framework. Therefore, apart from institutional constraints involving the need for an agreement, there are important motivation constraints that affect one side of our division of the Parties, namely, developed and industrialized nations. Moreover, the institutional constraints depend on the motivational constraints because, if we resolve the latter, the former will not pose a major problem for reaching an agreement based on rectificatory arguments. Hence, the relevant question now is whether motivational constraints (of the sort faced in the context of climate policies) suffice to conclude that a rectificatory approach is infeasible and, therefore, suffice to rule out the possibility of such an approach.

Interestingly, scholars raising the Political Feasibility Objection take the resistance of developed and industrialized countries to accepting rectificatory arguments as grounds for immediately rejecting a rectificatory framework. However, they do not seriously explore the possibilities of overcoming that resistance. In contrast, I follow Gilabert and Lawford-Smith in believing that motivational constraints do not immediately rule out the feasibility of a given option, but only force us to explore the possibilities of changing the motivational structures at stake. As they argue:

> The fact that people do not want to do something does not mean that we should think getting it done is infeasible, it just means we should think about how to change incentive structures and thereby change people's desires. In fact, political theories often function as social criticism, and,

> when they do, their main purpose is to help change people's desires and affect political apathy.
>
> (Gilabert and Lawford-Smith 2012, 5)

In the case at stake, the motivation of representatives of developed and industrialized nations, especially the United States, for not accepting rectificatory arguments depends, among other things, upon economic interests. An agreement based on historical responsibility is likely to harm their economic interests because they are among the biggest historical emitters. This fact is indeed difficult to change since this interest will remain unless other considerations outweigh it. However, in democratic political regimes, an important consideration is the democratic support that political actors receive from their populations. That is, the motivational structure of each nation's representatives is also affected by the motivational structure of their populations. If people support certain political initiatives, that can likely change their representative's motivation to accept them. For it is generally in the interest of political representatives to carry out and advance the initiatives supported by their (potential) voters.

These remarks suggest that the possibilities of changing existing incentive structures within the UNFCCC negotiations might depend significantly on the motivational structure of citizens living in developed and industrialized countries. Therefore, whether or not rectificatory arguments are likely to succeed in the negotiations on reparations for loss and damage depends partly on whether the general populace accepts those kinds of arguments.

A complete account of whether and how people can feel motivated to accept these kinds of arguments lies beyond the scope of this book, since that would require conducting an empirical study about the motivational effects of rectificatory arguments in the context of climate justice. My aim here is more modest. I rely on existing experimental studies and some background assumptions to suggest that people could actually be more motivated by rectificatory arguments than the Political Feasibility Objection suggests. If this is the case, then there are reasons to press these kinds of arguments despite current political actors' resistance to accepting them.

6.4.2 *Motivational and psychological aspects of the political feasibility objection*

In this section, I rely on selected psychological studies to suggest that rectificatory arguments might not be as infeasible as some scholars have believed. For, although they might currently be rejected by developed nations' representatives, they might help motivate their respective constituents to accept justice demands of reparations for loss and damage. If people are motivated by these arguments, this can have a mid-term effect on politicians' willingness to accept them. These findings weaken the case for the Political Feasibility Objection.

The argument I offer here relies on empirical data concerning people's motivation to rectify injustices. However, for these empirical data to support my argument, we must also rely on some background assumptions.[10] The first two assumptions concern conceptual relations. These conceptual relations involve identifying rectificatory arguments with arguments based on the breach of negative duties, on the one hand, and, on the other, forward-looking arguments with arguments based on positive duties.

I believe that these identifications are justified for the following reasons. First, rectificatory arguments relate to negative duties in the following way. Negative duties require that we refrain from harming others or from infringing other people's rights. As defined in this book, rectification aims to repair the consequences of the infringements of rights and the harm caused by climate change–related activities. Rectificatory claims and arguments emerge as a result of the breach of these negative duties. Second, forward-looking arguments relate to positive duties as follows. Forward-looking arguments aim exclusively at addressing people's suffering (in this case, suffering related to loss and damage from climate change) without delving into the question of how that suffering came about. Positive duties require assisting those who suffer harm, regardless of who has caused that harm. Thus, forward-looking arguments in the context of climate justice stem from positive duties to help those suffering climate change–related losses and damages.

My argument also draws on two further assumptions, following Lawford-Smith (2012). First, violations of negative duties are associated with actions, and second, violations of positive duties are associated with omissions. The justification for those assumptions is that, on the one hand, the suffering we want to correct through rectification stems from an action that violates negative duties. On the other hand, a failure to distribute resources in a way that meets the necessary standards to alleviate people's suffering is a failure to act in what we should do or should have done.

With these assumptions in mind, I argue, first, that rectificatory arguments might be successful in motivating people to repair losses and damages because people are more motivated to alleviate harm when they perceive that it has been caused by actions rather than by omissions. This phenomenon is known as 'omission bias' (Hauser, Tonnaer, and Cima 2009; Baron and Ritov 2004). For instance, in an experiment that tested this intuition, subjects faced two different trolley scenarios. In the first scenario, individuals were asked to consider the action of an individual pushing a fat man off a bridge to prevent the death of three people. In the second scenario, they were asked to consider a person omitting to pull a lever to stop the fall of a fat man that would prevent three people's death. Participants of the study judged the action in the first scenario as worse than the omission in the second (Cushman, Young, and Hauser 2006). Whether or not this bias is justified, this fact might be relevant when developing theories that can motivate people to act in ways we consider to be morally required.

These results support the claim that people might be more motivated to alleviate losses and damages from climate change when they understand them to be caused by actions rather than by omissions, since they will perceive those losses and damages as worse than if they had been caused by omissions. Arguably, other things being equal, the worse people consider suffering to be, the more motivated they might be to alleviate that suffering. Given that rectificatory claims emerge from the fact that certain actions cause human rights infringements, people might be more motivated to repair losses and damages when they are persuaded by rectificatory arguments. In my view, although this does not constitute a conclusive argument against the feasibility objection, it does speak against it.

My second point is that people might be more motivated to repair losses and damages when they are persuaded by rectificatory arguments because they are more motivated to repair harm when they believe that it is the result of their having breached their obligations toward others. This point is supported by another cluster of psychological studies, which have shown that people are more motivated to make up for the results of certain actions when they feel they have 'not fulfilled, or perhaps even . . . violated, one's own obligations toward others and society' (Caprara et al. 2001). The motivational force of that feeling has at least two manifestations. First, it motivates individuals to make reparations for the harm they have caused, and, second, it motivates people to refrain in the future from the actions that have caused that harmful situation (Smith and Lazarus 1990). However, these findings need to be qualified with two assumptions. First, we must assume that people are still more motivated to make up for the results of certain actions when they contribute to certain harmful effects together with others; and/or, second, that they are also motivated when they perceive themselves as part of collective entities causing that harm.[11] However, granting these assumptions, given that rectificatory arguments highlight the relevance of having breached the general obligation of not infringing people's rights, these arguments are more likely to trigger the necessary motivation both to repair losses and damages and to avoid contributing to actions that cause them in the future.[12]

In the previous section, I followed Gilabert and Lawford-Smith in arguing that the fact that people are not currently motivated to do something is not a decisive reason to believe that it is infeasible. It only means that we should think about how to motivate them to do it. I also argued that representatives of developed and industrialized nations might be motivated to accept a rectificatory framework for repairing loss and damage from climate change if the citizens they represent support that kind of framework. In this section, I have provided some experimental studies that, interpreted in certain ways, suggest that people might be motivated by rectificatory arguments based on historical responsibility. This suggests that rectificatory arguments, far from being unfeasible, might change political tendencies in the mid-term if those kinds of arguments are emphasized within political negotiations and within the public discussion.

6.5 Conclusion

This chapter has provided an overarching picture of rectificatory justice for loss and damage. In doing so, I have addressed some open questions. First, I have evaluated the adequacy- and fitness-for-purpose of attribution methods for identifying climate change–related harm for which liability would then be distributed in a policy mechanism for loss and damage. I have argued both the probabilistic approach and the storyline approach are adequate for the purpose of informing a rectificatory approach to loss and damage. However, I have also argued that the probabilistic approach is fitter for the purpose of liability because it reflects the same reasoning involved in legal liability practices, such as toxic tort law cases.

Second, I have argued that liabilities should be distributed within a policy mechanism for loss and damage in proportion to emissions, similar to the way in which a market share mechanism has been used to distribute liabilities in the context of legal liability for damages. Further, I have argued that rectificatory justice does not pertain to individual instances of loss and damage but rather to the structure of a policy mechanism designed to address these instances collectively.

Furthermore, I have argued that my account can meet the challenges brought against rectificatory justice accounts, such as those related to the time and effort victims might need to invest to achieve justice and those related to the lack of ability to foster adaptation. Against these objections, I have argued that my account based on a policy mechanism does provide a stable mechanism for victims to achieve justice without forcing them to go before the court. I have also argued that adaptation and loss and damage should be kept as two distinct policy responses to climate change, although synergies between them should be fostered.

Finally, I have addressed the Political Feasibility Objection. This objection asserts that a rectificatory- and responsibility-based account generates a polarized atmosphere that hinders consensus during political negotiations on climate policy. In order to answer this objection, I have explored the constraints that give rise to unfeasibility concerns and argued that they are highly dependent on people's incentive and motivational structure. Therefore, whether these constraints make the proposal unfeasible crucially depends on the possibilities for overcoming them. Next, I have explored the possibilities of overcoming these constraints by considering empirical findings concerning people's motivational structure. I have argued that some psychological studies may suggest that framing loss and damage as a matter of rectification may better motivate people to support policy mechanisms to address loss and damage. This suggests that if rectificatory arguments permeate from climate negotiations to the public discourse, they might actually change politicians' incentives to reject them. In my view, these remarks suggest that the Political Feasibility Objection may be less problematic than is often assumed.

Notes

1 A closed system is one that includes all the values obtained by application of a given operation to its members. As Oreskes et al. point out, only purely formal logical structures, such as proofs in symbolic logic and mathematics, can be shown to represent closed systems. That means they are the only verifiable systems because they can be proved by symbolic operations and the meaning of the elements is fixed and not contingent on empirical inputs (Oreskes, Shrader-Frechette, and Belitz 1994, 541).

2 The problem of deductive verification is that model results are always underdetermined by the available data (*underdetermination thesis*). In climate modeling, if the process fails, there is no simple way to know whether the principal hypothesis or some other auxiliary is at fault. If we compare the prediction of a model with the observational data and the computation is unfavorable, then we know that something is wrong, but we do not know what exactly. But even if the empirical data matches the prediction, there is no

way to know whether the hypothesis of the model (M1) constituted the necessary and sufficient conditions for the event (E) to occur or whether there is also another model (M2) that produces the same empirical data (E), so that some or even all of the elements of M1 were not necessary and sufficient conditions. It may also be the case that two or more errors in the model cancel each other out, so that a faulty model may appear to be correct. The upshot is that, when it comes to models of natural systems, it is simply impossible to isolate the necessary and sufficient conditions of an event. Therefore, verification cannot function as a benchmark for accepting climate models.

3 For detailed information about this case, see Mellor (2017).

4 Arguably, this argument would at least need to reasonably assume that there is no known factor which would have made the worker more prone to lung cancer than the general population, for example, a congenital health condition or some other factor such as smoking. The statement about exposure to asbestos doubling the risk must be based on an aggregate study of many individuals, so the implicit assumption is that the particular individual is not distinguishable from the population considered in the aggregate study. I thank Ted Shepherd for this point.

5 Similar legal cases where the threshold of double risk has been invoked can be found in Peñalver (1998) and Grossman (2003).

6 Arguably, the point here could be understood in two ways. This point could be understood in the context of climate litigation as addressing the likelihood of scientific causation proofs being accepted in courts. However, I do not intend it to be restricted to these contexts. My point is that the probabilistic approach better reflects our broader understanding of causation in liability contexts. Decisions made by courts in similar contexts serve as an illustration of the social understanding of the suitability of certain proofs of causation in those contexts.

7 Note that the probabilistic event attribution might be subjected to this problem when using proxy definitions that ignore dynamic changes (see Chapter 5).

8 While there might be different interpretations of who the relevant agents are, I rely here on Kenahan's proposal of considering the COP within the UNFCCC as the relevant agent. However, I will qualify this claim below.

9 For the sake of coherence, I have replaced 'j' with 'φ' in the quote.

10 In this section, I rely on Lawford-Smith's work (2012). There, she uses this psychological research to argue that cosmopolitan arguments based on 'justice' (to refrain from causing harm) are more likely to succeed than cosmopolitan arguments based on 'humanity' (to provide aid to those who are suffering). Here, I reframe her approach and apply it to rectificatory and forward-looking arguments. However, in essence, the distinction remains the same.

11 I do not claim that they are as motivated as they would be if they acted alone. This is called the diffusion of responsibility effect. However, I do assume that they are still somewhat more motivated than when they do not perceive the injustice as being caused by their actions at all. In my view, it is implausible to assume that by adding other people's actions to the equation, they would be less motivated than when they are not at all involved in causing the injustice.

12 Note that the same psychological research also suggests that guilt does not trigger feelings about the obligation to repair an injustice straightforwardly. It only triggers those feelings when individuals perceive remedial actions as an actual possibility. If people believe they have actual opportunities to make reparations, they are motivated to do so. However, if they feel they do not have actual opportunities to repair harms, they will instead be overwhelmed by fear of punishment (Caprara et al. 2001). However, this does not impact the argument of this book because, as argued in Chapter 7, rectificatory duties related to material reparations only apply to those who have the capacity to bear the costs of the negative effects of climate change.

References

Baron, Jonathan, and Ilana Ritov. 2004. 'Omission Bias, Individual Differences, and Normality'. *Organizational Behavior and Human Decision Processes* 94: 74–85.

Boran, Idil, and Joseph Heath. 2016. 'Attributing Weather Extremes to Climate Change and the Future of Adaptation Policy'. *Ethics, Policy & Environment* 19 (3): 239–55.

Brennan, Geoffrey, and Phillip Pettit. 2007. 'The Feasibility Issue'. In *The Oxford Handbook of Contemporary Philosophy*, edited by Frank Jackson and Michael Smith. Oxford; New York: Oxford University Press.

Caprara, Gian Vittorio, Claudio Barbaranelli, Concetta Pastorelli, Ivo Cermak, and Sandor Rosza. 2001. 'Facing Guilt: Role of Negative Affectivity, Need for Reparation, and Fear of Punishment in Leading to Prosocial Behaviour and Aggression'. *European Journal of Personality* 15 (3): 219–37.

Cushman, Fiery, Liane Young, and Marc Hauser. 2006. 'The Role of Conscious Reasoning and Intuition in Moral Judgment: Testing Three Principles of Harm'. *Psychological Science* 17 (12): 1082–89.

Gilabert, Pablo, and Holly Lawford-Smith. 2012. 'Political Feasibility: A Conceptual Exploration'. *Political Studies* 60 (4): 809–25.

Grossman, D. A. 2003. 'Warming up to a Not-so-Radical Idea: Tort-Based Climate Change Litigation'. *Colombia Journal of Environmental Law* 28: 1–61.

Hauser, Marc, Franca Tonnaer, and Maaike Cima. 2009. 'When Moral Intuitions Are Immune to the Law: A Case Study of Euthanasia and the Act-Omission Distinction in the Netherlands'. *Journal of Cognition and Culture* 9 (3): 149–69.

IPCC. 2021. 'Climate Change 2021: The Physical Science Basis'. In *Contribution of Working Group I to the Sixth Assessment Report of the Intergovernmental Panel on Climate Change*, edited by V. Masson-Delmotte, P. Zhai, A. Pirani, S. L. Connors, C. Péan, S. Berger, N. Caud, Y. Chen, L. Goldfarb, M. I. Gomis, M. Huang, K. Leitzell, E. Lonnoy, J. B. R. Matthews, T. K. Maycock, T. Waterfield, O. Yelekçi, R. Yu, and B. Zhou. Cambridge; New York: Cambridge University Press.

Kenehan, Sara. 2017. 'In the Name of Political Possibility: A New Proposal for Thinking About the Role and Relevance of Historical Greenhouse Gas Emissions'. In *Climate Change and Historical Emissions*, edited by Lukas Meyer and Pranay Sanklecha. Cambridge; New York: Cambridge University Press.

Lawford-Smith, Holly. 2012. 'The Motivation Question: Arguments from Justice and from Humanity'. *British Journal of Political Science* 42 (03): 661–78.

Lloyd, Elisabeth A., and Theodore G. Shepherd. 2021. 'Climate Change Attribution and Legal Contexts: Evidence and the Role of Storylines'. *Climatic Change* 167 (3–4): 28. https://doi.org/10.1007/s10584-021-03177-y.

Lusk, Greg, and Kevin C. Elliott. 2022. 'Non-Epistemic Values and Scientific Assessment: An Adequacy-for-Purpose View'. *European Journal for Philosophy of Science* 12 (2): 35. https://doi.org/10.1007/s13194-022-00458-w.

Mellor, Jonathan. 2017. 'Heneghan v Manchester Dry Docks Ltd: The Limits of the Fairchild Exception'. *Legal Issues Journal* 5: 129.

Moellendorf, Darrel. 2014. *The Moral Challenge of Dangerous Climate Change: Values, Poverty, and Policy*. New York: Cambridge University Press.

Oreskes, Naomi, Kristin Shrader-Frechette, and Kenneth Belitz. 1994. 'Verification, Validation, and Confirmation of Numerical Models in the Earth Sciences'. *Science* 263 (5147): 641–46.

Otto, Friederike E. L. 2023. 'Attribution of Extreme Events to Climate Change'. *Annual Review of Environment and Resources* 48: 813–28. https://doi.org/10.1146/annurev-environ-112621-083538.

Page, Edward A. 2012. 'Give It up for Climate Change: A Defence of the Beneficiary Pays Principle'. *International Theory* 4 (02): 300–330.

Parker, Wendy. 2009. 'Confirmation and Adequacy-for-Purpose in Climate Modelling'. *Proceedings of the Aristotelian Society, Supplementary Volumes* 83: 233–49.

———. 2020. 'Model Evaluation: An Adequacy-for-Purpose View'. *Philosophy of Science* 87 (3): 457–77.

Peñalver, Eduardo M. 1998. 'Acts of God or Toxic Torts? Applying Tort Principles to the Problem of Climate Change'. *Cornell Law Faculty Publications. Paper* 730: 40.

Posner, Eric A., and David A. Weisbach. 2010. *Climate Change Justice*. Princeton: Princeton University Press.

Priest, George L. 2010. 'Market Share Liability in Personal Injury and Public Nuisance Litigation: An Economic Analysis'. *Supreme Court Economic Review* 18 (February): 109–33. https://doi.org/10.1086/659983.

Rawls, John. 1999. *The Law of Peoples; with, The Idea of Public Reason Revisited*. Cambridge: Harvard University Press.

———. 2001. *Justice as Fairness: A Restatement*. Cambridge: Harvard University Press.

Sheffet, Mary Jane. 1983. 'Market Share Liability: A New Doctrine of Causation in Product Liability'. *Journal of Marketing* 47 (1): 35–43. https://doi.org/10.1177/002224298304700104.

Smith, C. A., and R. S. Lazarus. 1990. 'Emotion and Adaptation'. In *Handbook of Personality: Theory and Research*, edited by L. A. Pervin, 609–37. New York: The Guilford Press.

Stott, Peter A., David J. Karoly, and Francis W. Zwiers. 2017. 'Is the Choice of Statistical Paradigm Critical in Extreme Event Attribution Studies'? *Climatic Change* 144 (2): 143–50. https://doi.org/10.1007/s10584-017-2049-2.

UNFCCC. 2012. 'Ad Hoc Working Group on Long-Term Cooperative Action under the Convention (2012, August 15)'. *Report on the Workshop on Equitable Access to Sustainable Development. Doha.*

Wallimann-Helmer, Ivo, Lukas Meyer, Kian Mintz-Woo, Thomas Schinko, and Olivia Serdeczny. 2018. 'The Ethical Challenges in the Context of Climate Loss and Damage'. In *Loss and Damage from Climate Change. Concepts, Methods and Policy Options*, edited by R. Mechler, I. Bouwer, T. Schinko, S. Surminski, and J. Linnerooth-Bayer. Cham: Springer Publishing.

Final conclusions

Climate change is not a distant reality anymore. People, especially those most vulnerable, already suffer the negative effects of climate change in ways that negatively impact the core elements of their well-being and their human rights. Those impacts constitute loss and damage from climate change. At the same time, we know that climate change does not result from the natural variability of the climate. Instead, it is a reality arising from forms of production and consumption based on the massive use of fossil fuels. Given these two considerations (namely, the severity of the impacts and the origins of climate change involving human agency), I have characterized climate injustice in terms of human rights infringements, and I have argued that this injustice gives rise to rectificatory claims, which should be addressed by those closely connected to the injustice. In this book, I have developed a rectificatory framework for loss and damage based on direct historical responsibility for pollution.

Robert Musil wrote in 'The Man without Qualities': 'it is never decisive what one does, but always what one does afterwards'. This statement has been the *leitmotiv* of the thesis presented in this book. Even if industrialized countries did not know or could not avoid infringing human rights through their participation in climate change, now they have the opportunity to make up for the negative results of their pollution on the core features of people's lives. They have not only the opportunity but also the duty, because avoiding human rights infringements is such a weighty reason that cannot be easily outweighed by other considerations, especially for those countries that enjoy high levels of wellbeing. This is the first core message this book attempted to convey.

The second core message of this book is captured by Greta Thunberg's famous words to the US Congress: 'Listen to the scientists!' However, in this case, this is meant as a friendly reminder to my dearest moral and political philosophers working on climate justice: we can do so much better in shaping our theories of justice in light of new scientific developments and particularly in the case of attribution science. Although I acknowledge the difficulty of this task, I believe it is worthwhile not only because this will contribute to the accuracy of climate justice claims but also because it is a way of empowering vulnerable communities in their fight against climate change. The more scientific support their demands have, the stronger their claims will be. By no means has this book done all that can be done to live up to the highest standards of this call. However, I believe this is a good way to start thinking about how attribution science can help rectificatory justice claims.

Climate negotiations and the development of L&D policies within the UNFCCC have operated as the background for this book and motivated the policy relevance of this work. The development of a policy mechanism to address loss and damage has been a demand of the least developed and most vulnerable countries since the run-up of the UNFCCC in the 90s. Despite the advances toward a Loss and Damage Fund both in the Paris Agreement in 2015 and at the COP27 in Doha in 2022, some key issues remain unaddressed: the independence of L&D from adaptation measures, what the fund will finance (i.e., what will count as loss and damage and what measures are necessary to address them), who will be the donors to this fund (i.e., how will loss and damage duties be distributed), and how we will identify where loss and damage occur.

This book has contributed to addressing these remaining issues in various ways. First, it justified the independence of a loss and damage mechanism from adaptation measures by singling out the idiosyncrasy of loss and damage. Second, it provided an account of what it ought to count minimally as loss and damage and thus should be covered by the fund. These two contributions are presented in Chapter 2, where I developed an account of what constitutes, at least minimally, loss and damage and why L&D should be kept separate from adaptation. The chapter proposed what I called a minimal capability-based account of loss and damage, according to which loss and damage occur, *at least*, when climate change disrupts people's lives by pushing them below a sufficient standard in their opportunity to enjoy the central aspects of a dignified, flourishing life (or capabilities) and when such a disruption constitutes an infringement of their human rights. Based on that, I argued that L&D measures are ex post measures that apply when human rights have already been infringed, in comparison to ex ante mitigation and adaptation measures. The chapter also provided a categorization of each type of loss and damage (from economic to non-economic loss and damage) and offered reparative measures to address different manifestations of loss and damage.

Third, this book provided a justification for a direct principle of historical responsibility to identify donors and distribute burdens among them. This contribution is achieved in Chapters 3 and 4. Chapter 3 concluded that the PPP does not have the disadvantages the literature has identified and thus should not be ruled out in favor of the alternative rectificatory justice principle, the BPP. After that, Chapter 4 provided a justification for the PPP based on the Continuity Account. This account followed the spirit of Musil's words above and argues that polluters should bear the duties of addressing loss and damage because those duties stem from a previously unsatisfied duty of not infringing human rights. The chapter explained how this account can circumvent the main objections against a direct principle of historical responsibility and how it improves existing accounts.

Fourth, this book developed an account of how scientific findings can be used to identify loss and damage in Chapters 5 and 6. Chapter 5 presented and discussed the main attribution methods to link environmental loss and damage with climate change: the probabilistic approach and the storyline approach. It showed that the choice between both attribution methods should not be based on the storyline approach overstating the effects of climate change. Chapter 6 argued, in a nonconclusive way, that even though those two attribution methods are adequate for the purpose of identifying loss and damage from climate change, the probabilistic approach presents some more affinities with legal reasoning in liability contexts that make it fitter for the purposes of building a rectificatory policy mechanism for loss and damage. This chapter also explains how a policy mechanism for loss and damage could function following the input provided by attribution studies and how liability could be distributed once the results from attribution methods are considered.

Index

attribution: assessments 89; claims 93, 101n4; conditional 85; culpability 51; duties 23m2, 25n34; event 88, 95, 101n5; harmful effects 87, 90; method(s) 22, 85–86, 100, 109, 112–114, 117, 128, 134–5; purposes 112;question 88, 97–8, 103n20; responsibility 14, 71, 77; science (of) 6, 86, 89, 99, 115, 120, 134; statements 94, 99, 102n14, 120; study (pl.) 50, 87–88, 96–7; 95, 98, 100n1, 112–114, 117, 120, 135
agent-centered 42–45

beneficiary 49–53
beneficiaries 49–60, 63n12, 82n2
beneficiary pays principle (or BPP) 49–61, 61n1–n15, 82n2, 109, 134

capabilities 2, 12, 21, 29–45, 68–9, 109, 134; *see also* sufficiency threshold (of)
causation 53–9, 62n4, 94, 112, 114–19, 130n6; *see also* Causation Objection
Causation Objection 21, 49–60, 62n4, 62n12, 66, 84–86, 102n12, 109; *see also* causation
climate models 88, 95–100, 110–1, 119
compensation 3–10, 22n5, 23n15–n18, 63n17, 66, 93
Continuity Account 21, 66–82, 13

dynamics 87, 89–92, 96–100, 103n19; *see also* thermodynamics

ex ante **34**–36, 44–5, 134
excusable ignorance objection 49–61, 62n6–n7, 62n12, 63n15, 71–82
ex post **34**–37, **44**–45, 134

fairness 55–61, 63n15, 74, 85
feasibility: political 22, 121–129

harm 7–14, 18–21, 24n21, 25n32, 29–33, 37, 42, 43, 44, 50–60, 61n3, 62n4–n8, 74, 80, 84, 113–128, 130n10–n12; counterfactual notion of 30

isolationism 11–13; isolationist 11–12

justice: claims 8, 19, 59, 71, 85, 133–4; climate 2, 5, 10–11, 13, 18–22, 25n35, 28, 49, 56, 62n8–n10, 66, 81, 82n1, 84–5, 121–22, 127; commitment (to) 18; compensatory 7–8, 123; considerations of 12, 60; corrective 8; currency of 12; distributive 8, 12, 19–22, 25n32–n35, 33, 69; demands (of) 9, 20, 21, 69, 121, 126–7; duties of 24n26; duty-bearers of 14–16, 19; intergenerational 7, 17; loss and damage (for) 2, 66, 109–10; principles of 6, 13, 49, 53, 60, 81–2; primary subject of 121; procedural 78, 82; rectificatory 2, 7–10, 19–22, 25n32, 25n35, 69, 109–123, 128–9, 134–5; restorative 7; scope of 3; theories of 3

legal: cases 114, 130n5; contexts 93, 114, 118; domain 112–4; grounds 93; perspective 112; practice 114, 120; purposes 101; reasoning 22, 135; regime 23n18; scheme 57; standards 9; system(s) 56, 63n17, 74, 112–3; thinking 119; *see also* liability
liability 5–7, 22, 23n18, 62n6, 63n16–n17, 86–7, 93, 100, 109–129; contexts 22, 86–7, 115, 130n6; counterfactual

75–78; fitness-for 115, 118; strict 11, 51, 56–61, 63n12, 73–75, 135
life 29–31, **32**, 33, 36, **40**, 41, 44, 56–61; decent 12, 19, 20; disruptions 29–34, 69; flourishing 21, 29–32
loss and damage 2–22, 28–45, 66, 70, 75, 80–2, 84–86, 101, 109–114, 119–129, 133–135

model: computer 111; climate 87–8, 94–100, 102n15–n18, 110, 112, 119, 130n2; developers 111; evaluation 110–11
minimal 2, 21, 25n32, 29–33, 134

non-identity 30–1

overstatement(s) 22, 85, 89–100, 101n8–n20

PEA community 87–100, 101n4, 101n11, 113; *see also* probabilistic approach
policy mechanism 2, 3, 22, 85, 109–115, 119–122, 134–35
polluter(s) 13, 18–21, 23n16, 25n34, 42, 49–50, 56, 66–82, 92, 109, 113, 121, 133–134; *see also* Polluter Pays Principle (or PPP)
polluter pays principle (or PPP) 3, 6, 13, 25n31, 49–61, 62n6, 63n12, 66–82, 109, 120–1, 134; *see also* polluter(s)
probabilistic approach 21, 22, 85–89, 91–100, 102n15, 103n23, 113–128, 130n6, 134; *see also* PEA community

reparations 2, 7–10, 16–18, 23n19, **34**, 37–45, 69, 75, 81, 112, 120–129, 130n12; symbolic 9, 16–18, **34**, 41–45
responsibility 14, 16, 17, 28, 43, 62n4, 66, 70–3, 77, 129, 130n11, 133; historical 6, 7, 15, 21–2, 68, 73, 81–2, 110, 113, 122–8, 134; outcome 78–81, 82n3
risk(s) 1–5, 35–6, 41, 44, 54–6, 75, 78–80, 86, 89, 91, 100n2; intolerable 35–6; management 35–6, 45n2

state(s) 5–7, 12–20, 41, 43
sufficiency: level(s) 33, 36, 37, 42, 45, 69; *see also* sufficientarianism; sufficientarian
sufficientarianism 14, 19; *see also* sufficiency; sufficientarian
sufficientarian 13, 20–1, 25n31–n35, 32, 33, 69; threshold notion 30–33; *see also* sufficiency; sufficientarianism
storyline approach 21–2, 85–100, 101n5, 101n7, 102n14, 109, 112–120, 128

thermodynamics 89–92, 96, 99–100, 103n19; *see also* dynamics
threshold(s) 20, 25n32–n35, 32–3, 43, 87, 90, 93, 100, 101n4, 102n12, 111, 130n5; capabilities (of) 34, 44; dignified life (of) 30; harm 19, 31, 33; justice (of) 14; meteorological 94; moral 68; normative 8; notion of harm 14, 30–1; sufficiency (of) 2, 20, 21, 25n35, 30, **34**, **44**, 69; sufficientarian 33; well-being (of) 13
transgenerational: entity 15–19; groups 24n27

uncertainty (pl.) 95–99, 102n16–n17, 119–20

victim-centered 41–45, 53

Warsaw International Mechanism (or WIM) 4–6, 23n18

FSC
www.fsc.org
MIX
Papier | Fördert
gute Waldnutzung
FSC® C083411

Kolibri 360 GmbH
Ferdinand-Jühlke-Straße 7
99095 Erfurt, Deutschland
produktsicherheit@kolibri360.de